Plant Environment of Man between 6000 and 2000 B.C. in Bulgaria

Plant Environment of Man between 6000 and 2000 B.C. in Bulgaria

Tzvetana Popova

BAR International Series 2064
2010

Published in 2016 by
BAR Publishing, Oxford

BAR International Series 2064

Plant Environment of Man between 6000 and 2000 B.C. in Bulgaria

ISBN 978 1 4073 0470 0

BAR Publishing is the trading name of British Archaeological Reports (Oxford) Ltd.
British Archaeological Reports was first incorporated in 1974 to publish the BAR
Series, International and British. In 1992 Hadrian Books Ltd became part of the BAR
group. This volume was originally published by John and Erica Hedges Ltd. in
conjunction with British Archaeological Reports (Oxford) Ltd / Hadrian Books Ltd,
the Series principal publisher, in 2010. This present volume is published by BAR
Publishing, 2016.

Printed in England

BAR
PUBLISHING

BAR titles are available from:

        BAR Publishing
        122 Banbury Rd, Oxford, OX2 7BP, UK
EMAIL   info@barpublishing.com
PHONE +44 (0)1865 310431
FAX   +44 (0)1865 316916
        www.barpublishing.com

Contents

PREFACE

This study discusses the essential results of archaeobotanical studies carried in the territory of Bulgaria over the last five years with a special focus on the archaeobotanical finds from 36 prehistoric sites from the period Neolithic–Bronze Age. In addition the author presents a review of the history of the Bulgarian archaeobotany and its achievements. The scope of the analyses is considerable. For each of the sites are provided physicogeographical characteristics, the dating of the sites and the leading archaeologists. In order to obtain palaeobotanical (palaeovegetation and palaeoenvironmental) evidence contemporary methods and analyses have been applied (contextual and tafonomic) to the found herbaceous and ligneous plants for both the cultivated and the weed plants. For each of the studied settlements seeds and fruits as well as charred wood have been identified and examined.

In the discussion are analyzed the data for each of the sites and objects and the origin, the ecological requirements of the particular type, its spread in the different epochs – and the reasons of their finding are explained. In most of the studies are also described the wild growing species of plants, whose seeds and/or fruits provide additional information about the eco-climatic conditions in the respective settlements in the respective periods.

The determination of the charred wood also provides additional information about the composition of the forest vegetation in the respective region.

Most of the settlements have been situated at low altitude and that explains the wood of the Austrian pine (Pinus nigra) spread in the past at the low lands and now almost absent. Quite often the charred wood is from different species of oak, ash tree, horn-beam, alder, common English elm, maple, etc.

The basic types of cultivated and wild growing plans are being examined. Often they have been recovered from archaeological contexts and they played essential role in man feeding. Plants which were additional sources of food or were used for different technical needs were also examined.

The significant results of this study were compared with similar results from other archaeological sites in Bulgaria as well as from archaeological sites in neighboring territories.

The studies of the 36 sites provide valuable new data regarding the following three regions – the valley of the rivers Struma and Mesta (Southwestern Bulgaria), the Southeastern part of Bulgaria and Northern Bulgaria. Several new species such as chick peas and stone-pine were examined for the first time in this territory.

The cultivated cereals were einkorn, emmer and barley and also Cornelian cherry, blackberry, raspberry were usually collected. Palaeobotanical analyses of this study support the great diversity of cultivated, wild growing and weed plants in this region and also provide evidence for the availability of Mediterranean species plants such as cultivated vine and stone-pine which could have been transferred and distributed from Greece or Asia Minor.

ACKNOWLEDGEMENTS

This book is a result of my deepest links with the ancient plant remains and colleagues in Bulgaria. I am grateful to many individuals for their help.

My first thanks go to my original advisor Prof. E. Bojilova, Sofia University "St. Kliment Ohridski" for the important discussions and advices during the work on this book. Special thanks are due to Ass. Prof. Dr. Spasimir Tonkov, Department of Botany of Sofia University "St. Kliment Ohridski" for his reference.

I am most grateful to all colleagues who were generous with giving me their archaeobotanical materials and who made possible this study and especially to St. Alexandrov, Iv. Panajotov, Y. Bojadzhiev. I owe a further debt to the undergraduate, graduate, and doctoral students such as Nikolina Nikolova, Geny Vassileva, Lubomir Angelov, Ilia Hadgipetkov etc. from the Department of Archaeology of Sofia University "St. Kliment Ohridski" who endured bags of soil. Special thanks are due to Vania Petrova and Nadezhda Todorova from the Department of Archaeology of Sofia University "St. Kliment Ohridski" for a valuable discussion regarding the Archaeological characteristics of the regions.

Two other persons were of special help – Goro Katsarov for the graphic work on the map and Ivan Vajsov who gave me a permission to use his map. I am grateful also to Ass. Prof. Dr. Krasimir Leshtakov for his support and patience allowing me to revise my text during the time of the excavations.

I hope I have adequately acknowledged my indebtedness to the work of following friends and colleagues without whom the release of this book would have been impossible: Alya Veder who finalized the text translation on time despite of her family difficulties; Morena Stefanova, Research Associate in the Metropolitan Museum of Art for her suggestions, and critical queries who edited and improved the most part of the text; Georgi Ivanov who provided this crucial steps in technical preparation of the manuscript for publication, D-r John Tydeman for his advise.

Finally, I wish to thank to the publisher for his incredible patience and to my family for their support and understanding.

Chapter 1

INTRODUCTION

The present study aims to present the most significant results of archaeobotanical studies carried in the territory of Bulgaria covering the period from Neolithic to Bronze Age. The study is based on the analyses of archaeobotanical plant remains recovered from 36 prehistoric sites.

My main goals are:

— To investigate and define the main species of domesticated and wild growing plants which have had significant role in the nutrition of man and which appear most often in archaeological excavations.
— To characterize the weed flora which is of special significance for studying of ancient agriculture and the interactions of man with its environment.
— To study the charred wood remains and to make an attempt at palaeoecological reconstruction as well as to define the wood species which have been the most popular for everyday use.
— To comparatively analyze the results of my studies with those of other studied prehistoric sites in Bulgaria as well as with those of other sites from the neighboring territories in analogical periods.
— To define the basic cultural and wild growing plant species that had decisive role in the nutrition of man and appear most often in archaeological excavations.
— To discuss the tendencies in the archaeobotanical work, concerning the application of contemporary methods (contextual and taphomonic analyses) for defining the nature of agricultural process and the possibility of its reconstruction.
— To discuss the contemporary models and concepts connected with the economy of the ancient societies.

The archaeobotanical study is closely related to the study of the archaeological sites. The study of the flora and fauna as well as the way of interaction of ancient people with the natural environment is "a subject of environmental archaeology" (Wilkinson and Stevans, 2003). The study of the organisms in the archaeology is called "bioarcheology"

and the archaeobotanical investigation of charred remains which is related to the result of man activities Wilkinson and Stevans (2003) called "ecofacts". The "Environmental archaeology" has rapidly grown in significance and it is now seen as a major component to most excavation projects studying the human impact on the natural environment. It covers the period from the appearance of contemporary man ca. 40 000.

"Environmental archaeology" adds new data to those achieved as result of the archaeological studies and contributes to revealing details of the everyday life of man as well as for defining what have been the flora and fauna of the site and what kind of natural resources then man has made use of.

The interdisciplinary study of the origin and dissemination of the cultural plants with their accompanying vegetation adds data about the human communities development to that what has been revealed through the archeological study.

The development of the Bulgarian archaeobotany started at the beginning of the twentieth century. Its founder was Prof. Arnaudov published during the period 1936 – 1955 his studies of plant remains from 16 archaeological sites dated from different epochs. In the mid-seventies of the twentieth century the Bulgarian archaeobotanical studies has taken a next step associated mostly with the work of foreign scholars - Hopf (1973) who studied plant remains of 12 settlements, Jane Renfrew (1979) who, on the basis of previously published data from several settlements drew out conclusions about the use of the basic cereals cultivated in the Neolithic and Chalcolithic epochs, Robin Dennel (1974, 1978) who analyzed the materials of the settlements located by Chavdar village, the town of Kazanlak and the village of Ezero, Eva Haynalova (1975, 1980) who studied the materials of the settlement by the town of Nova Zagora and the village of Golyamo Delchevo, Yanushevich (1983) who studied part of the charred plant remains of the layers of late Neolithic period (5000 B.C.) from the settlement near the village of Ovcharovo – 'Gorata' near the town of Targovishte, Behre, K. (1977) who studied the Chalcolithic settlement located by Sava village near the town of Varna. After this period the studies of archaeobotanic plant remains in Bulgaria have been carried out further on basically by Bulgarian scientists, though often with joint participation in Bulgarian and international expeditions for collection of materials.

The scientific research in this third period in the development of the Bulgarian archaeobotany has been already characterized by special attention given to precise application of the stratigraphic method in the Bulgarian archaeology. Here should be mentioned the published results of archaeobotanical studies of prominent Bulgarian scholars such as Chakalova, Bojilova, Popova, Marinova, Sarbinska and Tonkov, Bojilova and Chakalova, (1980), Chakalova and Bojilova (1981), Bojilova (1986), Popova (1990, 1991, 1995, 1999, 1999a, 2001, 2003, 2006), Popova and Bojilova (1998), Marinova (2001, 2002a, b) Marinova and Chakalova, (2002), etc.).

Especially valuable work is the bibliography of Licitsina and Philipovich (1980) which systematized the archaeobotanic plant remains from Bulgaria, Greece and (former) Yugoslavia.

Depending on the scope of the archaeological activity and on the different intensity of the archeological studies as related to time and place, not all geographic zones of Bulgaria nor all cultures of the periods are equally studied.

The materials subject to study from the middle of 1980s were basically extracted through flotation done by horizons and stratigraphic columns or they were studied by complexes as for example houses, ovens, ceramic concentration, floors, pits and other archaeological contexts.

Chapter 2

LOCALITIES AND CHRONOLOGY

The archaeobotanical studies are based on the materials collected from 36 settlements in the territory of the country (fig. 1). Each settlement is dated, with given location and geographical characteristic.

The archeobotanical material is studied by periods as the exact culture is still not defined for some of the settlements and, on the other hand, the material in some of them originates from a locality with still not defined or not precisely defined culture by the archeologist – so the culture is given only there where it is known.

The Neolithic in Bulgaria is divided into three stages: Early Neolithic– ca. 6000-5450cal. B.C., Middle Neolithic– ca. 6000-5450cal. B.C., and Late Neolithic – 4900 cal.B.C. The next period – Chalcolithic or Copper Age in Bulgaria corresponds with the period ca. 4900-3800 B.C., divided into early and late Chalcolithic. During the Neolithic, Chalcolithic and Bronze Age different cultural groups developed in the different regions of the country (fig. 2).

Opinions concerning the chronology and the periodization are given in the study by the regions. Certain differences are observed in the different kind of the settlements. In the region of Thrace were formed settlements of the type "settlement mount" or tell settlements. In Western Bulgaria the Neolithic tell settlements are in rather small numbers, while the so-called "open air settlements" are predominant for Northern Bulgaria. In the different regions of the territory of the country different cultures have developed and spread and significant number of studies is connected with the chronology of the Bronze Age.

The periodization scheme of the Bronze Age is initially proposed by Mikov in 1971, who takes for its beginning 2500 B.C., later Katincharov proposes for its beginning 2750 B.C. and Panayotov later accepts it to be 3500 B.C. The chronologic extension of the Middle Bronze Age is not disputed – 2000/1900 – 1600/1500 B.C. The Late Bronze Age is dated 1600/1500 - 1500/1200/1100 B.C. It covers the following cultures: in North-Western Bulgaria the group Baley- Ursoe, in Northern Bulgaria – influence of the cultures Verbiçoara and Tey, in North-Eastern Bulgaria – culture Koslojeni – Noa-Sabotinovka, in South-Eastern Bulgaria and Thracia – group Razkopanitsa VII – Assenovec as well as the so-called group Zimnitsa – Plovdiv – Çherkovna, materials of which are also found in Northern and North-Western Bulgaria (Boyadjiev, 2003, c.20,21).

According to the current periodization the Bronze Age is divided into three stages which were supported by the radiocarbon data:

— EBA – 3250/3200 - 2600/2500 B.C.
— MBA – 2600/2500 - 2200/2000 B.C.
— LBA – 1600/1500 - 1200/1100 B.C. (Boyadjiev, 2003).

The data of Boyadjiev show significant modifications on the basis of the radio carbonic dating which, according to him forces a reformulation of the recent chronology (Boyadjiev, 2003). Significant part of the studied settlements are tell sites and most of them are located in the Thrace such as Galabovo, Madrets, Mednikarovo, Yunatsite, Karanovo, Nova Zagora, Nebet tepe, etc, and the others are 'open air settlements' such as Kovachevo, Ohoden, Orlitsa, Varhari.

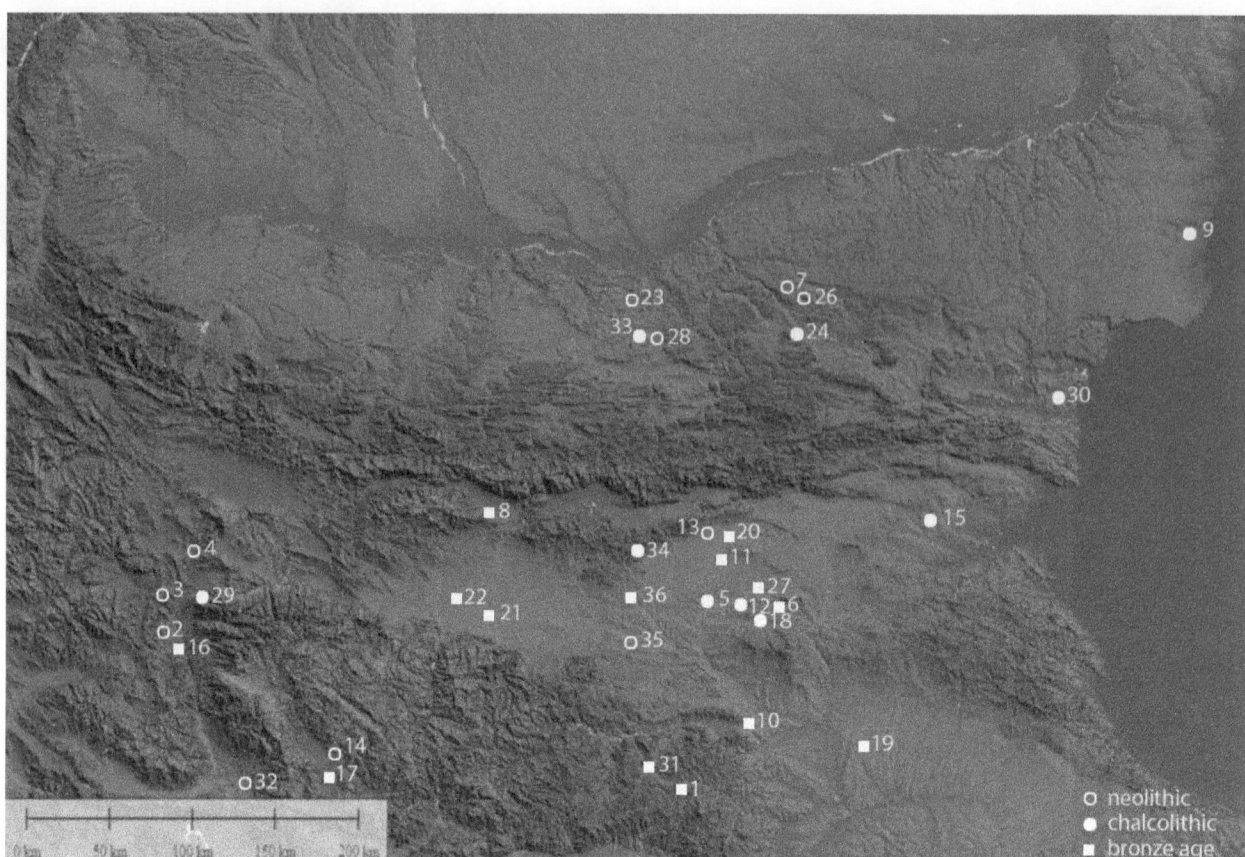

Fig. 1. Localities and chronology of the studied settlements in the territory of Bulgaria.
1. Adata, 2. Balgarchevo, 3. Vaxsevo, 4. Galabnik, 5. Galabovo, 6. Goljama detelina, 7. Drinovo, 8. Dabene,
9. Durankulak, 10. Dositeevo, 11. Djadovo, 12. Iskritza, 13. Karanovo, 14. Kovachevo, 15. Kozareva mogila,
16. Kamenska chuca, 17. Koprivlen, 18. Madretz, 19. Michalich, 20. Nova Zagora, 21. Nebet tepe, 22. Junatzite,
23. Orlovetz, 24. Omurtag, 25. Orlitza, 26. Podgoritza, 27. Polski Gradetz, 28. Samovodene, 29. Slatino, 30. Suvorovo,
31. Tatul, 32. Topolnitza, 33. Hotnitza, 34. Chatalka, 35. Jabalkovo, 36. Jazdach.

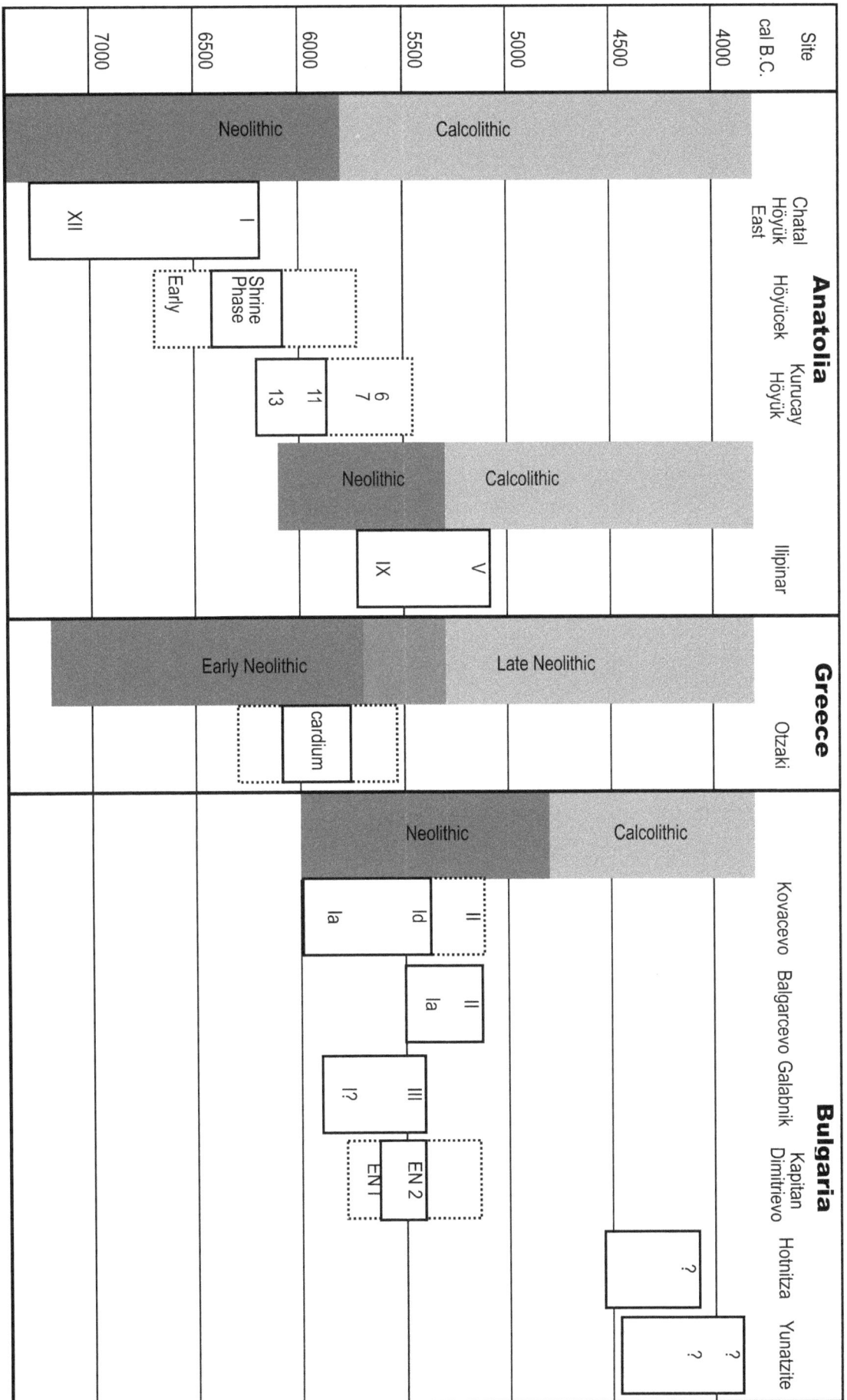

Fig. 2. Chronological settings of the sites mentioned in the text (Radiocarbon data after Görsdorf, Bojadziev 1996).

Chapter 3

METHODS AND ANALYSES

The development of the archaeobotany in Bulgaria could be divided into three stages during which different scientific methods were used. Prof. Arnaudov made a significant contribution in the understanding of the systematic of the species hulled wheat, leguminous, hulled and naked barley as well as in the consideration of the problems connected with the origin, dissemination and usage of the cultivated plants in Bulgaria (Arnaudov, 1937-1938; 1940-1941). Though the methods of study he used were still not the contemporary methods, nevertheless, his work established the solid basis of the archaeobotanical studies in Bulgaria, which started in the middle of the seventies of the twentieth century and, as it was already mentioned in the methodology, was connected mainly with the activities of foreign scientists in Bulgaria where most attention was given to some methodological problems (Hopf, 1973) as for example the connection between the level of carbonization of the wheat and the changes in their morphologic form. In the studies carried by Robin Dennel (Dennel, 1972, 1974, 1976, 1978) for first time new methods (such as stratigrafic collection of the materials and flotation) were used in the study and reconstruction of plant remains. The materials were collected from exact stratigraphic context by horizons and extracted through flotation. These methods made possible for Dennel to do the interpretation not only of the diversity of the species cultural plants but also to give his opinion about the agricultural economy of the settlements from which the materials originated. Thus, the work of Dennel was also methodological contribution to the correct interpretation of plant remains in different archaeological context and it appeared as the beginning of the implementation of the contextual analysis in Bulgaria.

The archaeobotanical studies in the contemporary, third stage of the development of the archaeobotany in Bulgaria are characterized with attention given to the exact stratigraphic methodology. A good methodology is established not only for work on the materials but as well for the way the materials are collected. In most cases the applied method is the flotation with taken under consideration stratigraphic columns, pits and horizons.

FIELD AND LABORATORY METHOD

The most important plant remains from the archaeological excavations are the seeds, the fruits and the charred wood. In some cases also fragment of daub are found. Theoretically any part of the plant could be found as an ecofact. According to Wilkinson and Stevans (2003) Ecofacts are also the pollen, the mushrooms spores, mosses and ferns in the archaeological structures and, if their spectrum and quantity are in high degrees and appear as result of human activity, they are also satisfactory reliable. The method of collection of macro remains which is in use in Bulgaria is the flotation method, or washing.

The flotation is done through a system of sieves with diameter of their opening of the size of 1,1 – 1,5 mm. After the samples are extracted from the sediments, they get dried and sorted. When the conditions of storage are good the most species of charred seeds and grains look as the contemporary ones. Therefore they could be identified through comparative collections. Such collection comprises samples of all species from different sources. The collection covers: cultural plants, stones of fruits, wild growing grass and weed plants. Atlases are used for additional comparison. The analysis aims also to define as exact as possible to what extent the quantities and the allocation of the remains are influenced by characteristics specific for the plant and the sediments and to what extent it is due to human activity. It is necessary a critical reconstruction of the past to be done, on the basis of both knowledge about the existing connections between different species deposits, originating from archaeological excavations, and of the conditions of their origin and preservation. On the other hand many factors should be also considered as for example the ecological and social behavior of some plant species. The different agricultural techniques also have their consequences. The best method for reliably done study is considered the systematically taken samples from extensive archaeological excavations and their consideration with the archaeological structures. The association of the samples with already known archaeological characteristics

Fig. 3. Flotation, washing of soil samples.

also adds reliability to tracing different kinds of human influence as well as it makes easier the final evaluation of the findings. The reconstruction and the analysis of the ecofacts is a basic part of the study by the archaeological excavations. I have studied and identified the macro remains from 36 prehistorically settlements by means of taphonomic processes and by the way of their finding in different structures and contexts, considering also the possible usage of the plant, the contemporary vegetation, with comparison between different settlements, the usage of the charred wood, etc.

ARCHAEOBOTANIC ANALYSIS

The archaeobotanical remains are collected in different ways – for most of them was used flotation. Defined are more than 161 plant species. 25 of them are wild growing trees, 11 species are fruits and 18 are bushes. The wild growing grass, cultural and weed vegetation number 107 species, 55 of which are cultural plants.

The determination is done by different categories: wheat, leguminous, oleaginous and fibrous plants, fruits, wild growing plants, indeterminable seeds and indeterminable fragments. Each of these categories plant remains is fully described and compared with findings in the Balkan region. For the seeds are used both reference collection and atlases (Cappers, 2009, Schoch, H., Pawlik et al. 1988, Vilarias, J. 1992).

CHARRED WOOD ANALYSIS

The charred wood is defined by means of digital microscope –Keynce VHX -100. The use of such a microscope allows the structure of the charred wood to be directly observed. Studied are three anatomical plans – transversal, longitudinal-tangential and longitudinal radial. For definition of the wood are used specialized atlases and reference

collections (Gregus, P., 1955, Schweingruber, 1988).

CONTEXTUAL ANALYSIS

Contextual analysis has been done in each of the sites and its aim is to discuss where the different archaeobotanical remains were found and their archaeological context – pits, granaries, floors, etc.

TAPHONOMY AND TAPHONOMIC ANALYSIS

The taphonomy is the science for the processes of fossilization of organic remains.

The term is introduced in 1940 by the Russian scientist Ephremov and the name comes from Greek language where "taphos" means 'bury' and "nomos" - 'law'. The main goal of the taphonomy is to study the processes of depositing of remains in the biosphere and to restore the conditions for formation and the place of the fossils in the lithosphere. The contemporary archaeological studies refer to the term 'taphonomy' for explanation of the processes of fossilization of the biological remains. There are two aspects connected with the archaeological data: structural and depositing. The structures are what man constructs – dwellings, embankments, banks, pits.

The deposits appear later as a subsequent act later and they are connected with man activity – as for example storage of grain into pits or stones and they are results of indirect processes, intentional or accidental. The archaeobotanists study the deposits which contain as their main part biological remains.

These remains are of different categories and they depend on how they got into the place where they were found. The processes connected with the burning are of basic importance and the level of preservation of the remains depends on them.

In that connection Hubbard, 1980 classified the plant remains in the following way:

Class A – The plant material got burnt in the same place (location) where found. This category covers different kind of burnt storages and pits.

Here the connection between the context and the plant remains is very close.

Class B – The plant remains are result of a single event connected with burning and it is re-deposed accidentally or intentionally. The act of re-deposing evidences that it does not concern and it is not connected with the context where it is found.

Class C – The material appears the result of several different events connected with burning and/or different activities. This material is mixed with archaeological elements – stones, flint, ceramics, etc. The remains here have only slight reference to the context where they are found. In that connection Fagan and Decorse, 2003 provide some basic taphomonic characteristics linked with the archaeological contexts.

— Process of formation of the monuments
— Preservation of ritual and family value and their secondary use
— Intentional or accidental demolishing
— Conditions of preservation of organic and non-organic material

The study of the collected archaeobotanical materials from the studied settlements is in accordance with the classification of Hubbard, 1980 and anywhere where it was possible was done the taphomonic analysis.

For the scientific terminology of the wheat and leguminous I have used the traditional nomenclature of Zohary and Hopf, 2000. This system does not follow the contemporary subdivisions based on cyto-genetic features but it is generally used by the archaeobotanists. For the rest of the plant remains I used the nomenclature of the Bulgarian Flora. The abbreviation cf. from the Latin 'confere' means that the title of the genus was used because the exact belonging of the species was not specified.

Chapter 4

Contextual and Taphonomic Analysis
of the Studied Settlements

Early Neolithic

BALGARCHEVO

Location and Physicogeographical Characteristic of the Region

Dating, Excavation leader– L. Pernicheva

The settlement is located in the middle river terrace on the right bank of Struma River and about 10 km northwest from the town of Blagoevgrad.

According to the excavation leader L. Pernicheva, it is dated from the Early Neolithic. Ascertained are 2 building horizons (Pernicheva 1995, pp. 99-128).

The archaeobotanical analysis shows presence of mixture of *Triticum monococcum* and *Triticum dicoccum*. The material originates from grain storage. The grain is not cleared off glumes. What makes a special impression is that the grains are large and their dimensions are bigger that the normal ones which most probably evidences the presence of intentional selection. For the leguminous – big quantities of peas were found. As admixture together with them are ascertained also single seeds of other leguminous – lentils and bitter vetch. The location of the settlement near to the river was obviously favorable for farming of leguminous plants which need more watering. The weed flora is presented only by three species of weed - *Gallium sp., Verbena officinalis, Ajuga chamaepitys*. The last two are indicators for limy soil.

TAPHONOMY

The grains in the grain storage as said above are well preserved with larger dimensions that those commonly met, which gives evidence that they were selected as sowing seed, the more so as they were found together with glumes. Thus they could be referred to Class A – the plant material burnt in the same place where found. This category subsumes different kinds of deflagrated storages and granaries. In this case the connection between the context and the plant remains is very close.

VAXEVO

Location and Physicogeographical Characteristic of the Region

Dating, Excavation leader – St. Chohadjiev

The prehistoric settlement Vaxevo is located in 'Studena Voda' countryside – situated in the first non-flooded terrace on the left bank of Eleshnitsa River. The terrain is a wide terrace situated 550-554 m above sea level.

7 horizons are defined. The first layer comprises the I-rst and II-nd horizon of the Early Neolithic period, characterized by white-painted pottery. The second layer embraces III and IV horizons, belonging to the final phase of the Early Neolithic, characterized by brown-painted pottery. The third layer consists of three horizons – one from the Late Chalkolithic, one from the end of the Chalkolithic period and one from the Early Bronze Age according to Chohadjiev (2001, pp. 7-9).

CONTEXTUAL ANALYSIS

The charred plant remains are found in the I st building horizon, dated from the Early Neolithic, where a pithos was found. It was found in a pit together with ceramic fragments from white-painted pottery. The contents of the grain mixture are about 150 gr.

The hulled barley is dominant in the found mixture – *Hordeum vulgare var. vugare*. Here and there in the mixture are found also grains of emmer - *Triticum dicoccum* –7 grains, einkorn - *Triticum monococcum* –51 grains; rye - *Secale cereale* –5 grains; bread/durum wheat - *Triticum aesivo/durum* – 1 grain Popova (2004).

TAPHONOMY

Pithos from I-rst building horizon - Crops to be storaged in pithos is a general practice but the material comes to us only in rare cases due to the presence of rodents or due to decay or decomposure.

The presence of the grains of wheat accompanying the barley could be explained by the fact that they were already in the pithos from a previous storage or they appeared there acccidentally together with the barley. All grains are small sized and fractured, from which it could be supposed that they were waste mixture or the conditions of farming in the region were not fairly favorable. The materials are referred to Class A.

It should be taken into consideration that this settlement is located in the mid-high lands, in infertile soil and remote of water, which could be the reason for the inadequate maturing of the ears.

KOVACHEVO

Location and Physicogeographical Characteristic of the Region.

Dating, Excavations leaders – V. Nikolov, M. Lichardus, L. Pernicheva and others.

The settlement is located in one of the terraces of Pirinska Bistritsa River, about 20 km to the west of Struma River.

According to the V. Nikolov and M. Lichardus it is dated from the early phase of the European Neolithic period. The latest period there is dated: C14 : 6830 - 6760 B.P., and the dating of palaeomagnetism is: 5590 - 5410 B.C. (Nikolov (1999a, pp. 59-66).

CONTEXTUAL ANALYSIS

The archaeobotanical samples are based mostly on printings on daub. A comparatively big number of fragments from walls and floors of houses was studied. The results show dominant presence of einkorn – *Triticum monococcum*, followed by emmer – *Triticum dicoccum*, as well as availability of naked and hulled barley – *Hordeum vulgare var. vulgare*; *Hordeum vulgare var. nudum;* common millet – *Panicum muliaceum;* grass pea - *Lathyrus sativus* (Popova, 1992a).

The data published by Marinova (2002c, pp. 1-11) from the same settlement, studied later, show the domination of emmer - *Tr. dicoccum*. The presence of grass pea is also documented by her as well as three more species of pulses were found: lentils – *Lens culinaris var. microsperma*, pea - *Pisum sativum*, bitter vetch - *Vicia ervilia.*

The analysis of the wild growing flora includes the following species - *Setaria viridis, Bromus sp., Phleum phleoides/tenue*, which according to Marinova are typical for dry climate and erosion (Marinova, 2002c).

The contemporary studies of the settlement show the invasion of Mediterranean elements, clearly evidenced by the presence of fruit of *Pistacia terebinthus L.* Cornel-tree stones as well as seeds of wild growing grape have been also found. All those plants were collected in the vicinity of the settlement and they were in the contents of the wild growing local flora. Fragments of fruit stones evidenced the presence of *Prunus sp.*

YABALKOVO

Location and Physicogeographical Characteristic of the Region.

Dating, Excavation leader – K. Leshtakov

Yabalkovo – the 'Karabilyuk' area – is located in the central part of the Upper-Thracian plain, in the Maritsa River valley. The 'Karabilyuk' area is in the vicinity of the village with about 50 acreage of land. The terrain is part of the second river terrace of Maritsa River, slightly sloped in western- west-north direction towards the river bank. The topography to the south is characterized by hills with volcanic origin. A big spring to the east forms a small swamp. The soils are carbonate, colored in beige-ochre to reddish.

The rescue excavations started in 2000 as part of a big-scale project connected with studying the region from archaeological point of view. The excavations run for 5 years and nowadays still go on. The settlement according Leshtakov, et al., (2007) is dated from the Early Neolithic period.

CONTEXTUIAL ANALYSIS

The study is based on 123 flotation samples and 189 fragments of daubs which have been taken from the Northeast sector. The material was collected from the following squares: sq. I 31; I 32; I 36; I 37; I 38; I 39; I 44; I 45, as well as from H 27; H 39; H 40; H 46. Different archaeological structures were studded such as wall destructions, houses, several early Neolithic pits, pithoses, remains from houses.

The samples from the early Neolithic house show a broad variability of species.

The contents of a vessel from square F17 comprise einkorn, emmer, bread/durum wheat, common millet and lentils, as well as fragments of charred oak wood.

Samples collected around another vessel showed the presence of hulled barley. Insignificant quantity of *Rumex crispis* was ascertained in a ceramic vessel found in sq. J40/K40. The samples from a pithos in sq. H29/H30 showed the presence of several oak fragments and those from a pithos found in sq. H17 except oak fragments contained also several seeds of corn-cockle – *Agrostemma githago,* as well as grains of vetch - *Vicia ervilia* were ascertained in a pithos from sq. H18.

In the house frequently were found only charcoals referred to the following species of wood: *Quercus sp.*, oak, several fragments of maple - *Acer cf. campestris, Alnus sp.* - alder and *Populus sp.* – poplar.

What made a special impression was that most of the plant remains were found in some of the samples collected from different archaeological structures in sq. I36. Thus in the Northwestern corner of the square were found several grains of barley, common/durum wheat and several grains in a state of extremely bad conservation due to which they could not be identified. Found in the same square by the dismantling of part of an artifact (most probably a pithos) numerous fragments of daub were studied. Defined here were 3 imprints of *Bromus arvensis* L. as well as several semi-preserved fruits of acorns and cornel-tree.

Two samples were collected from sq. I 37, where the goal was to define the daub from an early Neolithic pit. About 50 fragments were analyzed and they showed imprints and semi-preserved parts of acorn fruit. The consistence of that "wall coating" but was extremely easily crushable and falling into pieces. Most probably it was a kind of organic material that was lately charred.

The samples from sq. I 39 were taken from the Early Neolithic pit No 2. The flotation showed single grains of einkorn, bread/durum wheat, as well as several seeds of weeds: *Polygonum aviculare, Agrostemma githago.*

Cultural plants, though in insignificant quantity, were also found in sq. H40 taken from a pit (with depth of 2, 31 m) – containing emmer and lentils.

TAPHONOMY

Most of the archaeological contexts show waste activities from accidentally fallen grains of different cereals and leguminous. By the observation the conservation of the wheat grains was ascertained as extremely poor – they were deformed, broken and some of them could even not be identified. In such a case the material could be referred to Class C – material that appears as result of several different events connected with burning and/or different human activities. This material is mixed with archaeological elements – stones, charcoals, pottery, etc. The remains

here have only slight connection with the context in which they were found.

The other group of contexts included the contents of vessels, where the presence of definite quantities of grains, which were not entirely burnt and thus stayed preserved, was ascertained. The seeds of the found weed should have been included during the harvest and also collected and storaged together with the wheat.

These materials provide precise information so they could be referred to Class A – the plant material burnt in the same location where it was found. In this case the relation between the context and the plant remains is very close.

The biggest quantity of imprints was found in a level of destructions under pottery fragments under an oven. A significant part of them is stems of wild growing plants. Quite a big quantity of imprints of *Bromus arvensis* L. – 61 in number and of *Poa annua*; single grains of barley and of einkorn show their presence in the clay dough. In general, here are the imprints of cultivated cereals, however, instead of that there are found also wild growing cereal ears. The dominant species here is the *Bromus arvensis* L.

WEEDS

The following weeds were found: *Chenopodium album, Poa annua, Polygonum aviculare, Agrostemma githago, Rumex acetosella, Bromus secalinus.*

In the most places as single grains in the samples is found the *Poa annua* L. It is an annual cereal plant developing sprouts. It is found as weed in the vegetable and fruit gardens, in the cultures with combined surface, row crops, diluted sown fields of Lucerne, clover. Its typical habitat is the ruderal meadows and pasture-grounds of all kinds of soil. It develops successfully from early spring till late autumn. It propagates through seeds, partially vegetative. By availability of moisture in the soil the seeds germinate almost through the whole vegetation period – till late autumn. Its development is fast. It could be met also in the dark fallow land.

Chenopodium album L. The plant is a wide spread weed in the whole country. The young leaves could be used as food instead of spinach.

Some of the found species such as the goose-grass (*Galium apparine* L.), the *Rumex acetosella* L., the corn-cockle (*Agrostemma githago* L.) and the knotgrass *(Polygonum aviculare* L.) are met in the fields at the heights of the cultural plants, i.e. in the so-called 'second' level. These are the most widely spread species. They are typical for almost all fields and thus during the harvest they get into the sheaves. These weed could be of interest also as admixture

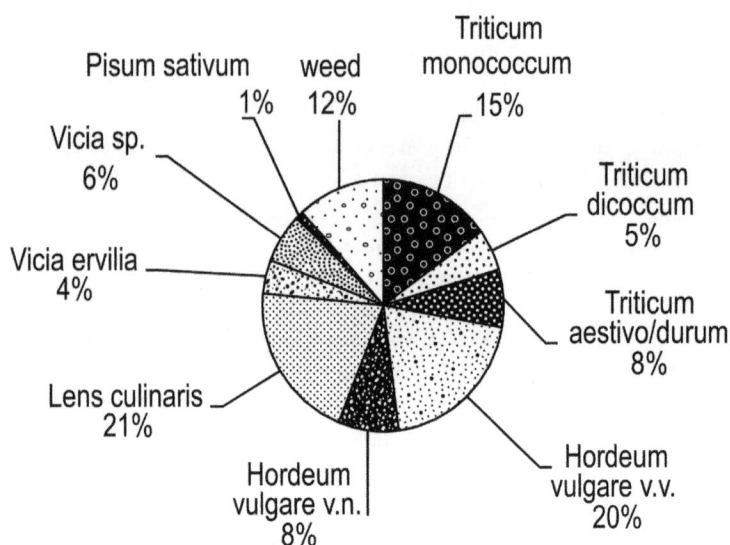

Fig. 4. Quantitative distribution of carbonized grains in the tell settlement Jabalkovo.

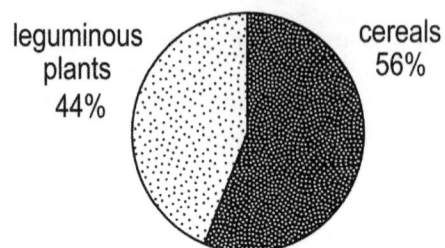

Fig. 5. Quantitative distribution of the cereals and leguminous plants in the tell settlement Jabalkovo.

to the seeds of the cultural plants. Here are the corn-cockle, the cleavers, the sweet grass, the clambering, etc.

Another groups of the studied plants are met in the dark fallow land, and namely: *Poa annua, Rumex acetosella*. Most of those plants are ruderal and they follow the human habitat. They have been widely spread in the country in the past and are until today. Some of them hold particular qualities which, most probably have been noticed and used. The wide spread of these species in ploughed and grass lands should have eased their collection and later usage. Chenopodium plants can be used to make soups and the young leaves are readily eaten by domestic animals.

Presence of characteristic weeds – *Chenopodium album, Polygonum aviculare, P. lapatifolium, Rumex acetosella* (found in the pits) can indicate spring sowing. *Chenopodium album* can be a common weed of spring-sown cereals. In spring the usual sowings are of barley and common millet, oats and from the leguminous – lentils and peas. The plant grows on humus and nitrogen rich soils as well as or clay soils.

Taking into consideration though the high percentage of leguminous, the dominant place of the barley and the availability of common millet, found in the context as well as the found weed, it could be guessed with some cautiousness that there were spring sowings.

CHARRED WOOD

A total 76 fragments of charcoal were found. Dominating are those of oak. Here and there in some samples: sq. I 36 were found single fragments of beech tree (*Fagus*); sq. I 45 - pit No 1 - Austrian pine tree; sq. I 37 - from an Early

Neolithic pit – fragments of pine and oak.

The studied fragments of charred wood provide the opportunity some conclusions to be done from economy point of view. The collection of woods was an everyday necessity and it took place in a not so wide scope of land. It is obvious that the lower and more easily accessible parts were preferred. Generally the data from the charred wood give evidence for the domination of oak forests with the presence by the river banks of alder trees, elm-trees, ash-trees and fruit-bearing trees. The data about the Austrian pine tree show its wider spread in the past compared to nowadays.

GATHERING

The data about it are documented by several fruits of cornel-tree and acorns.

OHODEN – 'Valoga' countryside

Location and Physicogeographical Characteristic of the Region.

Dating, Excavation leader – G. Ganezovsky

The 'Valoga' countryside ('Dolnite Laki') is situated 1, 5 km northeast from the village of Ohoden. The site is a flat one-layer settlement located in the slightly sloped to the east alluvial-meadow non-flooded terrace in the left bank of Skat River. Its area is of 10 decares. Uncovered is the eastern half of sunken structures, probably a house. It is dated from the Early Neolithic. The studied structures as well as the pits give good reason to the archeologist Ganezovsky (2008, pp. 20-35) to associate them with the second phase of Proto Starchevo culture.

I was provided with the archeobotanical materials of this site, thanks to the kind cooperation of the Ganezovsky, for what I express my special gratitude.

The samples are collected through flotation method. After the laboratory analysis were established plant remains of charred wood and seeds.

The samples though are extremely poor.

CONTEXTUAL ANALYSIS

The material is extracted from the following contexts:

1. PK L17, 4 center, structure No 4, depth 0, 15
2. structure No 4 in sq. K17, depth 0, 45
3. sq. L16/3-4
4. grave No 3
5. pit No 15 and No 16
6. 'dark brown stain' in sq. I 16.4

In the context PK L17, 4 center, structure No 4, depth 0, 15 were found two fragments from Austrian pine, and one fragment of oak.

From structure No 4 in sq. K 17, depth 0, 45 – three fragments of oak; in sq. L16/3-4 there is more diversity – here are found single fragments of oak, Austrian pine and two fragments of naked barley. A single seed of sorrel was found as representative of the weed.

Due to the still unclear significance of those structure till recently we could refer them with some reserve to Class C – The material is result of several different events connected with burning and/or different activities. The remains here have only slight relation to the context where they were found.

As concerning the wood we could confirm the availability of oak as well as of Austian pine. But the quantity of the fragments is quite insufficient at this stage for more precise conclusions about the local environment.

The data of the next contexts show the following:

Context No 1 – from grave No 3 – samples are taken from different depths. In some places there are traces of charring but the fragments are too small. Still we succeeded to establish the following species of plants: barley (because of the massive destruction of the caryopsis it is impossible to define if the barley is hulled or naked) – it is represented by one very poorly preserved fragment and only one whole grain. Ascertained is emmer, well preserved but with extremely small for its species seeds and a single seed of

mountain spinach. The material could be referred to Class A – The plant material in the same place where found.

Context No 2 – it covers pits No 15 and No 16 – in both pits was found oak. Its quantity in pit No 15 is quite significant. The fragments are large; they reach up to 10 sm. In the same pit are also defined several not charred fragments of plum but most probably it is a contemporary sort. Here also as above the material could be referred to Class A.

Context No 3 – 'Dark brown stain with coal char' in sq. I 16.4 – here two fragments of oak and one fragment of pine sort were ascertained. Because of the still not cleared significance of this structure it will be not discussed for now.

In general a concrete characteristic for the moment could not be done either for the cultivated plants or for the past environment. It is known that the different species of oak are widely spread, easily accessible and due to it they were used everywhere in different ways.

The finding of mountain spinach is not something unusual as the plant is a widely spread weed in the whole country whose young leaves could be used as food instead of the spinach. From this plant red paint could be obtained. And the seeds could be mixed with the animal fodder.

NEOLITHIC PERIOD

DRINOVO

Location and Physicogeographical Characteristic of the Region.

Dating, Excavation leader – I. Angelova

The site is located in the area 'Rezervata' in the lands of Drinovo village, in the district of the town of Popovo. The excavations were carried out in the period 1985-1988. The settlement consists of a cultural layer of the Early Neolithic period ('Ovcharovo' culture), one of the Early Chalcolithic period ('Polyanitsa' culture) and there are materials from the Late Bronze Age. According to Angelova (1998, pp. 91-96) the Neolithic materials have parallels in Ovcharovo – 'Gorata' (the Forest) and in Podgoritsa near Targovishte, Samovodene, Gradeshnitsa – Malo Pole and Karanovo II.

CONTEXTUAL ANALYSIS

The basic material for study from this site are daub fragments. They are collected following layers of Neolithic, Chalcolithic and Bronze Age. After the study of the material 175 clear imprints of cultural plants, cereals and leguminous, were identified. The following plant species were found: bread/durum wheat –*Triticum aestivo/durum,* as well as *Triticum dicoccum* Schrank. – emmer, which occupies the second dominant place as prelevance. The barley is also well represented. There are also imprints of the leguminous: *Lathyrus sativa* L. and bitter vetch - *Vicia ervilia* Willd.

LATE NEOLITHIC

KARANOVO

Location and Physicogeographical Characteristic of the Region.

Dating, Excavation leaders – V. Nikolov, St. Hiller

The tell settlement Karanovo is one of the biggest in South-eastern Europe. It is located in the village of Karanovo, at about 7 km north-west of the town of Nova Zagora, immediately to the south of the last ranges of Sredna Gora mountain. The tell settlement has an oval form with initial dimensions about 250 m to 180 m and a maximum depth of the cultural stratum 12,40 m. Part of its South and East periphery was destroyed by the local population, who used the tell as a career for production of construction dirt. The studies were started in 1936 by V. Mikov, later continued in 1946 also led by him, and in the period 1947-1957 V. Mikov worked together with G. Georgiev. The excavations were renewed as a joint Bulgarian-Austrian project by G. Georgiev and St. Hiller and since 1984 under the direction of V. Nikolov and St. Hiller till 2001 and continuinued at present by V. Nikolov (1995, pp. 23-24).

CONTEXTUAL ANALYSIS

The studied plant remains are collected by Prof. G. Georgiev from the floor of house No 2 from Karanovo III layer in the north-eastern sector of the tell settlement. According to the Nikolov (1995, pp.19-26) the house is dated in the first half of the Late Neolithic period in Thracia. 14 samples have been taken from this house. The quantity of each of them is different – from 80 to 150 gr. The extent of preservation of the grains is high which allows their determination as species. The following species are identified: *Triticum monococcum* - einkorn; *Triticum dicoccum* – emmer*; Lens culinaris* - lentils; *Vicia ervilia* – bitter vetch ; *Cornus mas* – cornel-tree; *Quercus sp.*- acorns; *Pinaceae* – coniferous; *Quercus sp.* - oak.

TAPHONOMY

Einkorn was found in two samples. The first of them presents mixture of einkorn and emmer. The second sample contains also mixture of those two wheat species but the einkorn is the dominant. Probably the mixing occured lately. These mixtures were purified from glumes and most likely they were ready for direct use.

Naked barley was found only in one of the samples. There it is found without any admixtures with other cereals.

Another sample contains bitter vetch with some mixture here and there of einkorn, which most probably appeared accidentally. The seeds of the bitter vetch from the both samples are with average dimensions and with typical for the species morphological indicators.

One of the samples presents sticked together heterogeneous frumentarous aggregation. Inside it were found fragments of and single grains of einkorn. The sticking together of the grains is due to water after which the carbonization took place. The crushed grains speak for a definite stage in their processing – most probably of grinding the grains. In one of the samples were found lentils seeds. The seeds were mixed with fragments of other corn species, here and there of einkorn. It is difficult to define the purpose of this mixture.

All species of seeds found in the house are cultivated and they are typical for the Neolithic farming.

GATHERING

The usage of wild growing resources is documented by fruits of cornel-tree and acorns which are extremely well preserved. The acorns are without cupula. And if they were used for the nutrition of the livestock it was not necessary to be cleaned from the cupula. On the other part the cupula get very easily separated from the fruit by burning when it is ripe. Thus it is very difficult to judge if they have not been used as supplement in the flour – as it was practiced by many tribes.

In the studied settlement were also collected also several burnt wood fragments. The defined species – oak and coniferous most probably were parts of the house construction.

Regardless that the materials belong to a closed complex (a house), having in mind that there many repeated activities took place there, we refer them to Class C - the material is the result of several different events related to burning and/or different activities. The remains here have but slight connection with the context by which they were found.

MIDDLE – LATE NEOLITHIC PERIOD

PODGORITSA

Location and Physicogeographical Characteristic of the Region.

Dating, Excavation leader – I. Angelova

The settlement is located to the north-east of the village with the same name on a plateau of average height. The studied lands cover an area of 6 dca. The collected materials according to the Angelova (1982a, pp.11-12) are dated from the beginning of the Late Neolithic (Usoe I culture). The findings and the pottery of the settlement have their analogies in the culture of the Neolithic settlements such as Karanovo II, Usoe I, Samovodene and Katchitsa I.

CONTEXTUAL ANALYSIS

By the excavations of the tell settlement Podgoritsa (Targovishte district) a big quantity of daub fragments were found .

The daubs fragments were collected from pit No 64 and they presented fragments of brown- reddish charred clay. On their surface and inside was found a big quantity of conserved plant imprints. 128 fragments with summarized volume of 5524 см³ were examined. All fragments were broken to pieces to find the imprints. Identified were imprints of grains, glumes, spikes, spikelet forks – which testify that plant admixtures, which represent the leavings left by the cleaning of the grains, were added to the clay dough. The species which were part of the composition presented in biggest number the hulled wheat – einkorn and emmer as well as hulled and naked barley.

On the basis of these data it could be suggested that in a Neolithic settlement the presented cultivated cultures were basically the hulled wheat and the barley. The presence of pea imprints provides evidence that there were also sowings of leguminous plants.

EARLY, MIDDLE AND LATE NEOLITHIC PERIOD

ORLOVETS

Location and Physic-Geographical Characteristic of the Region.

Dating, Excavation leader – P. Stanev

The village is located in the district of Veliko Tarnovo, in the central part of the Danube Plain. It spreads 9 km east of the river of Yantra and almost at the same distance from the Danube River and from the Middle Balkans. The terrain of the village Orlovets is hilly-flat with an average altitude of about 360 m. The soil belongs to the grey forest soil types. The potential vegetation in the area around included oak, lime-tree, yoke-elm, maple, manna-ash and others.

The Neolithic settlement Orlovets is located in the site 'Ada-kuzu' in the lands of the village of Orlovets. According to the archeologist P. Stanev (2008, pp. 24-27) it is dated in VII-VI millennium B.C. It was inhabited during the Earliest, the Middle and the Late Neolithic. Here are found houses, ovens, clay instruments of production and ceramic vessels. The conducted during 1993 archaeological studies found remains of a house. Findings are flour remains as well as with fragments of a single time remodeled oven. A new building level is defined from after a prolonged hiatus, where the remains of a new house are distinctly defined. Most probably this house was terrestrial but it was also destroyed by fire. There are no sufficient data about its structure and wall construction. Because of the insufficient information from the excavations of the house complex in Orlovets the archaeologist P. Stanev announces that neither the type of the houses nor the planning of the settlement could be more precisely defined.

CONTEXTUAL ANALYSIS

The archaeobotanical materials are collected from the I-st, II-nd and III-rd cultural horizons belonging to the Neolithic period. The quantity of the plant remains is comparatively small. The conservation of the grains is poor and part of them is deformed. Defined are some cultural plants among which hulled wheat – einkorn and emmer. The barley is presented by single grains which are also from the naked species. Found were also several grains of wild growing grapes.

TAPHONOMY

Probably all grains have got into here accidentally. The deformation of the grains and their poor conservation speaks for waste activity and shows different degree of carbonization which had impact on the process of deposition so they refer to Class C.

NEOLITHIC – THE END OF THE CHALCOLITHIC PERIOD

DABENE

Location and Physic-Geographical Characteristic of the Region.

Dating, Excavation leaders – L. Nikoilova, A. Bonev

The settlement is located 10 km south-west from the town of Karlovo. It is situated among the mountainous hills of the Balkans and Sredna Gora mountain chain. The local climate is continental with Meditteranean influence which benefits the cultivation of agricultural plants. The potential natural environment consists of oak forests that had significantly degraded so nowadays they are replaced by agricultural lands.

In the region were evidenced traces of Neolithic and Chalcolithic periods. Exactly from there originate also the first results taken by the initial excavations of the site.

The samples were collected by the leading archaeologist during the excavations. Unfortunately the archaeobotanical studies were limited only to the study of the provided material. Defined were imprints of emmer and bitter vetch. Studies of archaeobotanical remains from the Bronze Age were done lately by E. Marinova. She has defined the presence of the following plant species: *Triticum monococcum* L. – einkorn; *Triticum dicoccum* Scrank. (Schubl.) - emmer; *Triticum spelta* Godr. - spelt; *Triticum cf. aestivum* L. – common wheat; *Hordeum vulgare* L. – barley; *Lens culinaris* Medik. – lentils ; *Pisum sativum* L. – peas; *Vicia ervilia* Willd. – bitter vetch; *Linum isitatissum* L.- cultivated flax.

As elements of collection were found: *Cornus mas* L. – cornel-tree; *Prunus sp.* – plums; *Quercus sp.* - oak (acorns); *Sambucus ebulus* L.- elder, and from the wild growing and grass plants: *Bromus arvenis* L.; *Bromus cf. tectorum* L.; *Euphorbia cf. helioscopia* – milkweed/spurge; *Galium cf. spuruim* L. – lady's bedstraw/goosegrass; *Lapsana communis* L.; *Lithospermum arvense* L.; *Poa pratensis/tinalis*; *Vicia cf. tetrasperma* Moench. (Marinova, 2003)

MEDNIKAROVO

Location and Physic-Geographical Characteristic of the Region.

Dating, Excavation leaders – St. Alexandrov, I. Panayotov

The Neolithic settlement is of the 'open air' type and it is located in the periphery of the nowadays village with the same name situated on the high right bank of the Karapelitska River. As result of the seasonal floods caused by the river, in the Southern part of the site was formed a natural profile, where a cultural layer with a depth of about 3 m. was defined. The cultural layer is on a low natural elevation. Excavations were carried out by the archaeologists St. Alexandrov and I. Panayotov in 1993. An area of 130 sq. m. was studied. Differentiated are 4 levels of construction where minimal quantities of construction remains were found. According to V. Nikolov (1998, pp. 48-49) the pottery of the two lower horizons (IV and III) could be

referred to Karanovo II while those of the lower two buildings layers (II and i) refers to Karanovo III – IV.

Determined are V or VI building horizons, all of them from the period of the Neolithic. Later materials (Early and Late Chalcolithic) are documented only in a single trench, located at the edge of the terrace but outside of stratigraphical context. Panayotov and Alexandrov (1995), (Leshtakov, et al., 2001).

POLLEN ANALYSIS

The archaeological study based on the systematic archaeological excavations in this regions of the tell settlements Madretz and Galabovo and their vicinity involved implementing a pollen analysis for defining a more precise picture of the potential vegetation. The studied material originates from the trench No10, comprising archaeological findings dated from the Neolithic to the Early Bronze Age. The Neolithic samples were taken of the trench from each 20 cm down to 180 m depth. The samples were put to pollen analysis which was done in the Laboratory for palynology by the Biologic Faculty of the Sofia University 'St. Climent Ohridski'. After the chemical processing only 7 samples proved to be good enough for further analysis. From them about 300 pollen grains were extracted. Defined were several species of the family *Chenopodiaceae;, Poaceae, Compositeae* and from the trees – alder tree - *Alnus glutinosa*.

The found plants are typical representatives of the field and weed vegetation. Except them were also found several pollen grains of cultivated cereals but their more precise defining was difficult due to the not so good conservation of the grains. The found pollen grains of alder are a typical indicator of humid conditions but the quantity of the material is quite insufficient for defining concrete conclusions.

OMURTAG

Location and Physic-Geographical Characteristic of the Region.

Dating, Excavation leader – I. Angelova

The tell settlement is located 1 km to the north of the town of Omurtag. Its dimensions are around 3-4 m with a diameter of height – 0,7 m. Rescue excavations were done in an area of 2 dca. Four building horizons were defined. Three of them have been destroyed by fire. In the burnt horizon located in the Western part of the hill, in houses oriented to South-East and to South-West were revealed waste pits, filled with carbonized wheat, mortars, mill stones.

The settlement is dated from the Neolithic – end of the Chalcolithic – the second half of 5000 B.C. The collected material is referred to the culture Kodjadermen – Gumelnitsa – Karanovo VI Angelova (1982b, pp. 16-17).

A large quantities of carbonized remains as well as wall and floor daub were found. Most wheat mixtures were preserved in ceramic vessels.

CONTEXTUAL ANALYSIS

The material is collected from the vessels (horizon o, depth 0,40, square G5, s. 2, 10, p. 4, 70) where a big quantity of lentils was found. Its dimensions are as follows: length: 2,77 mm, width: 2,37 mm, i.e. it is with small grains and belongs to the sub-species *Lens culinarus v.microsperma*, the average dimensions of which are between 3, 0 and 6, 0 mm and *Lens esculenta v. мacrosperma* with average dimensions of about 6, 0 – 9, 0 mm. The lentils is 99,97% of the content and the rest are bitter vetch – 0,02% and grass pea – 0,09%.

TAPHONOPMY

The presence of the last two species is due to their accidental appearance or, as it is also possible, they were left in the ceramic vessels from a previous usage. They refer to Class A.

In another sample – taken again from a vessel – lentils seeds are 99,75% and as single seeds among the others are found seeds of bitter vetch. The seeds here are still smaller what makes us presume that it should be a semi-wild form with higher drought resistance. And that it was grown exactly for this reason. The einkorn – 88,19% dominates in another vessel, followed by the emmer – 5,08% and the bread/durum wheat – 3,43%. Small quantities of barley and durum wheat are also ascertained there.

TAPHONOMY

All seeds are cleaned from glumes. The mixture took place later. The seeds were found in vessel which ears that they were left for storage. The admixtures of the other plants of cereal are insignificant. They could be explained with accidental fall in the vessels or fall among during the collection and movement of the grain from one place to another. All materials refer to Class A – The plant material burnt in the same place where found and the relation between the context and the plant remains – very close.

The bitter vetch is presented with 28 seeds. The repetitiveness of the finds of this leguminous in North-Eastern Bulgaria indicates that it was surely used for nutrition of people and animals.

Except the carbonized seeds and grains were studied also fragments of wall and floor daub from the settlement. They are collected from the first layer of the central profile and dated from the end of Neolithic – the beginning of the Chalcolithic. Examined were 148 fragments with a total volume of 1837,75 cm³. Some quantity of them contained imprints of leaves, stems and grains. Quite often in the daub were found cereal imprints but more interesting were those of spelt and naked barley. The distinctive features of the spelt grains are: it is wider and flatter than the emmer. The correlation length to height is lower. The spire of the grain is blunt, widely curvaceous, the basis is slightly tapering and, as a whole the grain is more rounded than by the other hulled wheat species. The found imprints coincide in full with the charred material.

SUVOROVO

Location and Physicogeographical Characteristic of the Region.

Dating, Excavation leader – I. Ivanov

The settlement is located in the so-called 'Koriyata' neighborhood of the village with the same name. The existence of early Chalcolithic settlement is ascertained there. Studied are 450 sq.m. Defined are two buildings horizons – lower, from the Early Chalcolithic, and upper – from the Late Chalcolithic, according to Ivanov (1983, pp. 21-22). Only fragments of pottery originate from it as it is fully destroyed. Significant is the presence of the big pottery – storage. The settlement has been distorted by fire (Ivanov, 1983).

CONTEXTUAL ANALYSIS

The studied material presents plant remains collected around a pithos as well as fragments of daub. The total volume of the daub is 605,65 cm³.

Insignificant quantities of einkorn and emmer were found as admixture. No imprints were found on the pithos fragments. The daub fragments contained plant admixtures. Defined are glumes of hulled wheat, barley, imprints of stems and leaves of wheat plants. Found are also three imprints of grains from hulled barley and one of einkorn. A preliminary conclusion could be done that here in the period of the Chalcolithic period, were grown hulled wheat, hulled and naked barley.

CHATALKA

Location and Physicogeographical Characteristic of the Region.

Dating, Excavation leader – I. Dimitrov

The settlement is located in Stara Zagora district, 20 km to the west from the town of Stara Zagora. The excavations there were lead by I. Dimitrov. The settlement there existed from the early Chalcolithic period to the II-nd and III-rd phase of the late Chalcolithic period - the end of the culture Maritsa and Karanovo VI. The samples are provided by the archaeologist of Stara Zagora museum - I. Dimitrov. The quantity of the samples is comparatively low.

CONTEXTUAL ANALYSIS

Context: Ar. 13, hor. II b, depth 1, 60, north - 5, 00, east - 4,90 and it refers to the end of the early Chalcolithic period - the end of the culture Maritsa. It almost entirely consists of grains of naked barley – 94, 73%. As admixture appear also bread/durum wheat – 1,05% and durum wheat - 4, 21%.

TAPHONOMY

This admixture is due to the way of depositing of the material – from the floor of a house and thus it refers to Class C.

Context: 13, sq. A, building horizon IIb, depth 1, 70, north - 3, 00, east - 3, 00 – house. It consists entirely of einkorn. It is dated by the end of the Early Chalcolithic period – the end of Maritsa culture and refers to Class C.

Of special interest is the found caked mass of charred common millet. Single grains are well the mass is quite friable which is characteristic for the common millet.

Despite the fact that the quantity of samples is not so high, it is obvious that einkorn was presented in the settlement. A large quantity of hulled and naked barley and common millet was found. The common millet and durum wheat are presented only as single grains and it is difficult to say if wheat was grown on a mass scale.

VARHARI

Location and Physicogeographical Characteristic of the Region

Dating, Excavation leader – Y. Boyadjiev

The Late Chalcolithic settlement Varhari is situated in the lands of the nowadays village of Varhari, Momchilgrad area in the Eastern Rhodopes. It is localized at the place of confluence of the rivers Varbitsa and Diva Reka on an above-flood terrace slightly inclined from north to the south. At present the site is studied as a part of the archaeological survey conducting due to the construction of the road Djebel – Podkova. According to the archaeologists

Boyadjiev were revealed (Boyadjiev 2008, pp. 51-52) the site represent a settlement form the late Chalcolithic.

I was provided with the materials from the senior researcher Yavor Boyadjiev to whom I express my sincere gratitude. The samples comprise daub fragments and sediments with plant components. Their determination was done after their flotation.

CONTEXTUAL ANALYSIS

— Trench 13 north - 0, 50 m, west - 13, 70 m, depth from the surface – 0,50 m, - the sample is soil from the inside of a small vessel – no plant remains were found.
— Trench 42- 40- from ceramic heaping – several imprints of wild growing wheat were found - rye brom.
— Depth from the surface - 0, 30 m - 0, 70 m, and from 3 - 0, 30 m – 1,80 m. The sample contained 15 imprints – distinct imprints of einkorn (10) and 5 imprints of naked barley.
— Trench 13, central part, depth 0, 30 m - 6 fragments of daubs were studied and, correspondingly on them were found 13 imprints - 3 of einkorn and 10 of wild growing crops.
— Trench 12, depth 0, 80 m - 0, 57 m. The sample contains charred seeds of bitter vetch.
— Trench 13, west - 7, 80 m, north - 0, 20 m. Here again there are seeds of bitter vetch in the sample.
— Trench 13, west - 7, 60 m, north - 2, 87 m, depth - the same contents of bitter vetch.

Basically the samples are highly insufficient due to the character of the collection of the material. Although, on this site were carried only short-term rescue excavations we still could say that the bitter vetch dominates in the found materials and that they refer to Class A. In the imprints found in the daub dominate einkorn, barley and rye.

CHALCOLITHIC PERIOD

GALABOVO

Location and Physicogeographic Characteristic of the Region.

Dating, Excavation leaders – I. Panayotov, K. Leshtakov, St. Alexandrov

The tell settlement Galabovo, called 'Assara' is situated on the river terrace of Sazliyka River with a height of 100 m, above the sea level, located about 2 km to the east of the town of Galabovo.

CONTEXTUAL AND TAPHONOMIC ANALYSIS OF THE STUDIED SETTLEMENTS

The stratigraphy of the site according the archaeologists I. Panayotov, K. Lehsakov and St. Alexandrov consists of tree layers:

Layer A – not entirely preserved at the most part of the tell settlement, yielded materials from the Middle Ages, the Roman times, and the Iron Age.

Layer B – Early Middle Bronze Age.

Layer C – the end of the Chalcolithic period Panayotov, et al., (1991).

CONTEXTUAL ANALYSIS

All archeobotanical materials are extracted from layer C and include samples from I, II and III buildings horizons. The main part is collected from the I-st building horizon from the house in sq. No 7 and sq. No 8. A significant part of the material originates also from different archaeological contexts – ovens, houses, floor levels, hearths, millstone, pits, stratigraphic ditches, etc.

The studied material presents the following picture: the einkorn ranks first among the cultivated plants.

The emmer and the barley are presented in rather small quantities. From the hulled and naked wheat – the common and the durum are less often found.

From the leguminous is found the bitter vetch. It is frequently ascertained and in some places its quantity overcomes that of the wheat. It is the same when compared to other neighboring sites Popova (1992b), Popova (1994). Except it also daub were collected, whose fragments refer to the Chalcolithic layer. The number of the studied fragments is 192 and the results are similar by their species composition to those of the charred material.

TAPHONOMY

Of special interest are the significant concentrations of bitter vetch and lentils in vessels around the millstones. The found seeds are either separately found or in mixtures with wheat.

The question that appears is whether the bitter vetch has been milled as the wheat and used for flour. Because it is known that the plant is slightly toxically so it should be preliminary to cooking soaked in water. Thus, it becomes obvious that in the found admixture of bitter vetch and wheat, the bitter vetch should have been preliminarily soaked and only lately mixed for direct use. The materials of these contexts refer to Class A.

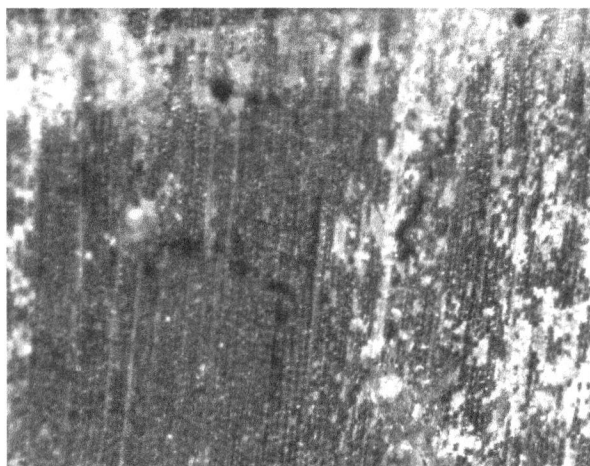

Fig. 6. Fossil wood from tell settlement Galabovo with deformation.

Charred wood has been found in 15 samples. It comprises all the three buildings horizons from the Middle Bronze Age. Defined are 286 fragments. The results, present the quality correlation towards the scrutinized samples by horizontals and for each of the species.

TAPHONOMY

In the settlement was found some quantity of wood with untypical characteristics which is the reason of drawing special attention in the study. The wood showed deformations which were not often observed by studying charred wood. The microscope observations resulted in several possible interpretations: the wood had been twisted before or after being charred. The way the cells are positioned, the structure and the deformation could be explained as result of high compression on the wood either it was deposited for a very long period in the sediments. In laboratory conditions such deformation could be achieved by extremely high temperature. Another opportunity is if the wood had degraded before been used by the prehistoric men. The wood is fossilized and it manifests effect of very high temperature. The analysis of this wood shows that it could be divided into two types:

A part of it has the characteristic features of re-deposited wood – as it had remained long in or had been carried along by water.

The other part is hard, fossilized and with shining surface. It is wood of juniper, fir-tree or cypress.

The data from the geologic studies in the region carried by Georgieva (1984, pp. 40-41) also define the contents of the located at the surface coal as originating from stems and branches of juniper.

The presented results of the study of the wood create the following picture: due to its wonderful qualities as construction material the oak had been the dominating species of plant in the region and, as such, used most in different domestic activities. In the contents of the forest wood was included elm-tree, maple, birch, hazel. In the hedges and in the outskirts of the settlement prevailed mountain ash, cornel-tree, cherries, plums and, most probably part of these plants (in wild or semi-cultural state) had been objects of gathering. The data from the anthracological analysis of the fruit trees could not always distinguish the cultivated from the wild growing species on the basis of their morphological indicators. That is why confirmations by a carpological analysis are needed.

The coniferous wood from cypresses and juniper shows specific deformations similar to fossilization which is characteristic for the coal alluvions in the region. These alluvions are not deep so most probably they were used for burning. The question which arises is how the pre-historic man knew about their existence and how did he use them. Possibly the coal was collected during certain periods of time. Wood with similar fossilization was found in the caves Cannaletes and Usclades in Massif Cental – France, dated from the Paleolithic (Thery, et all., 1995). For now but we still need some additional proof to confirm this fact.

The weed flora is represented by nine species. The correlation between the species in the different contexts is different. Impressive here is that the bigger part of the weeds is characteristic for the winter sowing - *Polygonum lapatifolium, Agrostema githago, Bromus secalinus, Rumex acetosella* and that they are annual plants. Taking into consideration that the lentils is sown in the spring, it is very possible these weeds to have grown whth it.

The dominant weed species in the whole assemblages were *Chenopodium album* L. and the *Polygonum aviculare*. It is known by literature sources that the epigeous parts of these plants are very good as fodder and their seeds are used in the nutrition of the birds. The wide spread of these plants in arable and grassy lands made their collection and parallel to it their usage easier.

Very often together with the cultivated plants in the archaeological contexts are also found plants from the wild growing flora. Such sorts have served as continuous resources and usually they were objects of intentional collection. The seeds and the stones are the most often found evidences as the soft parts of the plants either decay or burn. In the tell settlement Galabovo are found the following species of cultural fruit trees:

Fig - a typical species for the South Bulgarian Black Sea cost and widely spread there.

Grapes – documented both as wild and as cultivated species, intermediate forms. The issue of the grape is quite complex as its cultivation started very early but very often semi-cultural or outcast plants were grown in places.

Wild growing fruit trees: cornel-tree – one of most often met and widely used forest fruit. Its presence is confirmed also by the fragments of charred wood, so consequently it existed somewhere in the outskirts of the settlement.

White and black elder – the both species are often evidenced with presence in the outskirts of the settlements and they both are typical ruderal plants.

Oak fruits – The acorns of the oak were a source of nutrition and often during periods of famine they became replacement of the flour.

DURANKULAK

Location and Physicogeographic Characteristic of the Region.

Dating, Excavations - H. Todorova, Iv. Vayssov, T. Dimov, Y. Boyadjiev

The tell settlement Durankulak is situated on the Big Island in the Durankulak Lake. The archaeological excavations in Durankulak – started in 1974 – were carried out by an archaeological team directed by the senior researcher Prof. Dr. Henrieta Todorova.

The depth of the cultural layer is 3,50 – 4,00 m with eight stratigraphically divided building horizons. On the top was a Proto-Bulgarian settlement dated from IX-X century followed downwards by a layer of the Early Bronze Age and 6 layers from the Chalcolithic period. The archaeological materials refer to the cultures "Hamandjia", "Sava" and "Varna". Together with the Neolithic settlement, located on the western bank of the lake also a pre-historic necropolis (with 1 204 Neolithic and Chalcolithic graves) is thoroughly studied. The archaeological site shows a chronological continuity from 5 300 B.C. till the beginning of the XIth century. According to the archeological data there are ascertained processes of intensive cultural, technological and economical connections of the ancient population: to the north with the cultures "Boyan", "Bugo-Dnestrovska" and "Cucuteni-Tripolie". On the western bank of the Durankulak late are ascertained dug-out dwellings dated from 5 100 – 4 700 B.C., grave from the Proto-Bronze Age (3 500 – 4 700 B.C.) and late ancient Sarmatic necropolis. On the Big Island in the Durankulak lake is located also a Chalcolithic tell settlement (4 600 – 4 200 B.C.) defined as cultural heritage of national importance.

On the southern slope of the island there is a fortified settlement from the Late Bronze and the beginning of the

Early Iron Age (1300 – 1200 B.C.). 26 m inside in the rocky massive of the island a cave sanctuary of the Goddess Kibela form Hellenistic time (III century B.C.) was cut and on the entire island lands was located a fortified Proto-Bulgarian settlement from IX-X century A.C. (Todorova, et al., 1989); Todorova (2002).

Contextual Analysis

The materials are collected in the years 1981 – 1982 as well as sporadically in the years to follow. The samples originate both from the settlements and from the necropolises. The material refers to the Neolithic and Chalcolithic periods. The methods used are flotation and direct collection of material from hearts, graves and vessels. In addition large quantity of daub fragments is studied. Big part of the charred grains is destructed and mixed with fragments of charred wood and thus often it is extremely difficult the exact taxon of Triticum grains to be identified. In most cases they belong to wheat and barley. Einkorn and emmer as well as naked and hulled barley are found in a Neolithic layer. They are presented in the form of imprints where stems and straw with their characteristic morphological features could be distinctively seen. Except for these imprints there are also imprints with not so distinctive features that could be referred eventually to *Triticum monococcum/dicoccum* and *Triticum dicoccum/Triticum spelta* (Popova 1991a).

Taphonomy

Common millet is also found in the Neolithic layers. It presents a compact mass, easily broken and falling apart. As it was mentioned above, the compact status was resulted by the initial presence of water and the following fast process of charring. The glumes of the grain had preserved the characteristic shining for the common millet. The grain size and the form of the corn germa are the basis it to be referred to the species of common millet - *Panicum miliaceum*. (Presence of common millet was also defined in daub imprints). Single grains of bitter vetch were also found. They are small sized, with an average diameter of about 3 mm and they have a characteristic rounded-triangle form.

Several more plants were also used for nutrition. Such is the sorrel, found in a small vessel without admixture of any other seeds. These two samples are referred to Class A.

The availability of seeds of *Setaria sp.* and some pulses *Vicia* found on imprints could provide evidence for weeds in the sowings and especially in that of the common millet. The species *Setaria sp.* appear as a permanent satellite and accompanying weeds in the common millet. Such kind of 'specialization' has appeared as result of their similarity to the common millet for which reason they are difficult to distinguish in the process of growing.

In the Chalcolithic period the presence of the defined plants was almost identical. Thus, for example in a sample collected from the land between two dwellings were ascertained as presence grains of einkorn, emmer/spelt, lentils, and a single grain of *Polygonum aviculare*.

The quantity of the grains in the most cases is insignificant and the diversity of the materials comes to show that it is a secondary product – accidental split on the floor or results from different human activities, so they refer to Class C.

Quite bigger quantity represented also by several species of cereals was found in a house (in sq. L 121, building horizon VI). The charred remains were found in the granary. The emmer there is 66, 88% and the einkorn is 32, 15%. The rest in the percentage are single grains of durum wheat and barley. This mixture was cleared by the glumes and prepared for storage. According to how it was found it could be referred to Class A.

Additionally two stones of cornel-tree and seeds of black elder were recovered.

IZVOR

Location and Physicogeographic Characteristic of the Region.

Dating, Excavation leaders – St. Aleksandrov, K. Leshtakov

The site is situated North-West of the town of Radomir on the northern slope of the Konyavsko-Milevska Mountain range not far from the main road to Sofia (about 50 km far) and Kyustendil (about 39 km far). The village of Izvor is the biggest village in the southern part of the Radomir hollow. The biggest river in the lands of the village is Kosmatitsa River with main feeder Negomijska River. Kosmatitsa River further runs into Struma River. According to the archaeologists who lead excavations the site there refers to the Chalcolithic period.

Contextual Analysis

The archaeobotanical materials originate from trenches in the location 'Virovete'. The samples are collected from trench 1, trench 2, and 4 and present daub fragments. A total of 112 fragments were studied. The following imprints were found: common millet, barley, bread/durum wheat, peas. And from the wild growing - rye brom, prickly grass, vetch and common sorrel. The first two species are often met weed in the wheat.

ISKRITSA 2

Location and Physicogeographic Characteristic of the Region.

Dating, Excavation leaders – B. Borissov, K. Leshtakov

The site Iskritsa 2 is located about 100 m to the west from the nowadays village on the high left bank of Sokolitsa River. The studies here started in 1988 carried out by B. Borissov. The existence of a dated from the end of the Chalcolithc big open-air settlement was ascertained, that was destroyed by a big fire (Leshtakov, et al., 2001).

CONTEXTUAL ANALYSIS

The material presents daub fragments. They are collected by a Chalcolithic dwellings found in a ditch – 60 – 100 sm. 178 imprints were found, 94 of them belonging to einkorn, 45 of emmer, 38 of barley and 1 of bitter vetch. So the dominant plant here is the einkorn.

KOZAREVA MOGILA

Location and Physicogeographic Characteristic of the Region.

Dating, Excavation leader – P. Georgieva

The tell settlement is located in the lands of the town of Kableshkovo, Bourgas district. The excavations here were carried out by the archaeologist P. Georgieva. Found is a cultural layer from the Early and Late Chalcolithic as well as a thin and significantly destructed layer with materials from the Early Bronze Age.

The archaeobotanical materials were provided to me as result of the kind collaboration of the archaeologist Dr. P. Georgieva. The material originates from the Chalcolithic layer and according to Dr. P. Georgieva it refers to 'Varna' culture Georgieva (1999, pp. 415-425).

CONTEXTUAL ANALYSIS

The materials were collected by different archaeological contexts – around ovens, hearts, and burnt dwellings. They consist of charred grains and daubs fragments.

As result of the carried out analysis the belonging to different species of significant quantity of charred grains was defined as well as 49 exact daub imprints distinguished. The total quantity of the collected material is 6 kg, originating from sq. 2/23. The material contains many mineral admixtures but after sieving through a system of sieves

about 150 grams purified charred wheat grains were separated.

TAPHONOMY

The grains do not contain admixtures of other cereals species. All grains are very small, severely broken and fragile. From this sample a special test was done targeting objective measurement of the grains that defined that the grains were significantly smaller that the accepted as norm for the species. For that reason we accept that the collected grains present waste product in a certain stage of the grain processing as about 90% of the grains are in poor condition.

The emmer represents 1 kg of the samples. In the same way was done the sieving for extracting the admixtures and thus were defined 564 grains. The grains are whole and with good conservation. Among the main species of emmer were found also single grains of barley and several grains with intervenient traits between emmer and spelt. Most probably these grains appeared accidentally as their number is insignificant. The materials refer to Class B – the plant material is a result of a single-time activity and it had been redeposited accidentally or intentionally. The very act of redeposition shows that it does not specially refer to, neither is related to the context by which it was found.

MADRETS – Gudjova Mogila

Location and Physicogeographic Characteristic of the Region.

Dating, Excavation leaders – M. Dimitrov, K. Leshtakov, I. Panayotov

The tell settlement is located about 6 km to the north of the city of Stara Zagora. It is situated about 800 m far from the nowadays river bed of Sokolitsa River. Its diameter is 135 m north-south and 125 m west-east. Its height is 159, 6 m. The tell settlement was inhabited from the Early Chalcolithic period to the Early Bronze Age.

The first drillings were done by the archeologist M. Dimitrov in 1973. In 1992 Krassimir Leshtakov restarted the excavations and went further on in the next several years with some interruptions. In 1994-1998 the archaeologist I. Panayotov joined in the excavations. The data from the archaeological material allow the settlement to be dated from the Chalcolithic (Early and Late) and from the Early Bronze Age (EBA III). Layer B contains of 3 buildings horizons that could be referred to the culture Maritsa II-III. The latest building horizon refers to the culture Karanovo VI. Layer A contains 2 levels of inhabitance and the material could be referred to the Early Bronze Age – culture Sv.

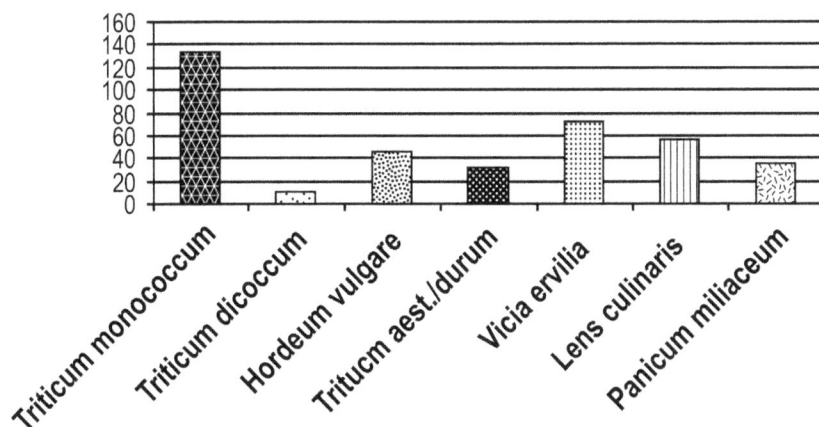

Fig. 7. Distribution of plants remains in heart 2 from tell settlement Madretz.

Kirilovo (Panajotov, et al., 1991; Leshtakov, et al., 2001).

From physicogeographical point of view the region is characterized with plain-hilled lay. Its climate is transitional continental. The alluvial gatherings are favorable for agricultural activities. Nowadays the region is strongly impacted by human activity – wheat, maize, tobacco, etc. are grown here.

CONTEXTUAL ANALYSIS

The samples are extracted through the method of flotation. The material was collected by the both layers. Stratigraphic study was done on the northern profile where samples were taken from in a continuity by every 10-15 cm from the depth of 63 cm to 385 cm.

The preservation of the material is not very good. The wooden fragments are with small dimensions which do not overcome 0, 5 mm. Another part of the material originates from ovens, hearths, and floors, concentrations of ceramic and stones. Daub were also collected for defining the imprints.

454 fragments of the charred wood were analyzed: 379 from the stratigraphic profile and 95 from the hearth.

ANALYSIS OF CHARRED WOOD:

Layer B – Chalcolithic (260-385 cm)

The analysis evidenced 10 taxons with the presence of oak in each of the samples. Species as *Corylus sp.; Carpinus sp.; cf. Betula; Alnus sp.* are found in a lower concentration. Wood of elm-tree, maple (*Acer campeste*) and *Rosaceae* is more often found.

ANALYSIS OF CHARRED SEEDS AND GRAINS:

Layer B – Early Chalcolithic

25 samples are collected altogether – 19 from the stratigraphical profile and 6 from different contexts – hearth No 2, pithos, pits, and floor. The following species were defined: *Triticum monococcum L. Triticum dicoccum Schrank.; Triticum cf. compactum Host.; Triticum cf. spelta L; Hordeum vulgare* though they are presented in low quantities.

The conservation of the grains is satisfactory. Still in their bigger part they are deformed and fragmented.

In the context – hearth in house No 2 the most abundant species are: einkorn, naked barley, lentils, bitter vetch. The results are summarized in a fig. 7.

Fig. 7 shows the presence and the abundance of the einkorn and other food plants.

TAPHONOMY

The diversity of the seeds shows that they are products of different separated activities, obviously of several cookings in the hearth and, accordingly, they have fallen aside accidentally as split material. They refer to Class C.

Around the ceramic concentrations were found the following species of plants: einkorn – *Triticum monococcum* L., emmer – *Triticum dicoccum* Schrank.; common wheat - *Triticum cf. compactum* Host.; wheat – spelt – *Triticum cf. spelta* L; hulled barley – *Hordeum vulgare var. vulgare* L.; lentils – *Lens culinaris var. microsperma* Mediк; bitter vetch - *Vicia ervilia* Willd.

By a lesser quantity are presented *Triticum aestivo/durum, Triticum cf. compactum; Lathyrus sativum.* The collected from the ceramic concentration grains of the defined plant species are with small dimensions, deformed and often broken – so obviously they appear secondary products.

In these samples, the seeds of *Agrostemma githago, Bromus secalinus; Rumex acetosella* were also find. The materials refer to Class C.

ORLITSA – Kirkovo Municipality

Location and Physicogeographical Characteristic of the Region.

Dating, Excavation leader - Y. Boyadjiev

The archaeological excavations are of rescue character. The site is located directly by the place of merging of the rivers Orlishka and Lozengradska, on the accessed by floods terrace in the right river bank. A settlement of 'scattered' type is defined, destroyed by fire. The houses are found at 2, 3 m depth. They are located at a big distance from each other. Remains of four houses were found. Most studied is a part of house No 1 which consists of two premises. The both are connected in their western ends by a construction of massive stones, glued with deep clay coating. It was destroyed by the fire and fallen down in pieces on the nearby situated vessels. From the northern house was cleaned only it's most southern part. The floor there was covered by floor-boards wide from 0,15 m to 0,2 m, with length of 4,1 m and with a depth of 3 to 4 m. The wall is made of stakes with diameter 8-10 m and in the middle of the wall there is a door wide 0,8 m. The western and southern walls of the southern room are made of wooden stakes with a diameter of about 10 cm, clayed and supported by arranged stones.

According to the excavation leader Dr. Y. Boyadjiev the site refers to the Chalcolithic period (Boyadjiev, 2004).

CONTEXTUAL ANALYSIS

The archaeobotanical materials were collected in the process of the rescue excavations. The method used was the flotation and also, wherever possible – done manually. The samples were collected explicitly from this dwelling as also from some other archaeological structures: in sq. D 1 and V1, daub from sq. D 5 in the same house, from sq. D 6 and E 4; from house No 5 – samples from it were taken from two pits – No 11 and No 12; from pit in house No 6 as well as from the following squares: C 27, D 32.

House No1 in sq. V1.

In this house was found a wooden floor of several big beams which were exposed to fire – with a length of about 5 m. The beams are extremely well preserved. Of special interest is the found near to them line formed by holes of stakes. Clearly defined were 20 holes with preserved wood in them. Samples were taken from each of these negative structures. The analysis of the charred wood of the stakes

showed the following: all stakes were done by oak wood. The average age (in years) of the trees had been 10, 6 years where the oldest tree used for the preparation of a stake was at about 20 years and the youngest – about 5 years. The average diameter of the woods is 9,7 cm, the width is 16 cm and of the narrowest one – 5 cm. The average depth of driving in is 4,8 cm, the deepest is 8 cm and the most shallow is 4 cm. The result of the analysis shows that a big part of the wooden material was from young trees at almost one and the same age. We could only suppose that they were cut in a single time from a single place.

In the same house were taken samples also from the beams. Two of them were from oak and the third one from pine tree.

The dominating role of the oak is confirmed also by the found in the holes wood remains in the floor at the southeastern corner of the dwelling as well as also from other contexts.

Of special interest for the scientific study is also the big number of fragments of ash-tree, extracted through flotation from the floor of the house.

The different species of oak are often found in the pre-mountainous lands in deciduous, broad-leaved forests and they develop well in rich and fertile soils by permanent humidity.

The second most often found species of plant in the studied settlement is the ash-tree. Its wood is extremely strong and it is especially good for elaboration of carpenter's tools. The species contents were enriched by the found fragments of hornbeam. The hornbeam is often found in the oak-beech forests. It grows well in fresh and friable-sandy soils. Its wood is good for the elaboration of small wooden objects.

Samples from sq. D 1

Of special interest is the found vessel full with bitter vetch seeds.

TAPHONOMY

In the vessel was found only bitter vetch. The grains are quite small. Single grains of pulses were marked also in some samples taken from Sector C. The materials refer to Class A.

The small dimensions of the wheat and leguminous as also the remains of wild growing plants in the wall daub lead us to the presumption that the weather conditions were unfavorable for the agriculture or that these findings were the products of crop failure.

House No 5.

In there are registered several beams located in its northern side as well as a series of holes of stakes. The beams are very well preserved. The wooden material in these contexts was proven to be again oak. In two places in the house is used coniferous wood – probably Australian pine – beam No 2 and beam No 3 by the northern wall. In a vessel from the same house – in sq. F 18 – was found wood from *Rosaceae.*

Concerning cultivated cereals in this house was found no such material. Marked were elements of collection – hazelnuts and cornel-tree in sq. H 18 and apple seeds in sq. F 18.

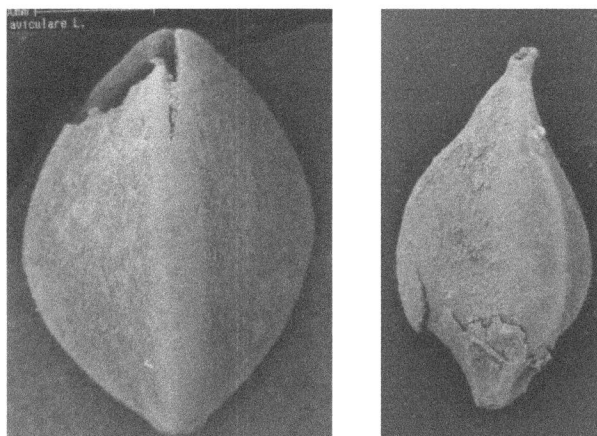

Fig. 8. Polygonum convonvulus - carbonized seed from Orlitza.

TAPHONOMY

Single grains of cereals and leguminous were found in sq. E 32 – in its northern part and in pit No 12 (II level) – bitter vetch, einkorn, lentils. In pit No12 was found only common millet. It is not possible to be established if their depositing was connected with secondary activities or the pit was used for storage because of the insignificant quantity of the found material and, correspondingly, due to the degree of charring. It is not very likely the pit to has been used as storage place so we refer the material to Class B.

On the basis of the restricted quantity of cultural plants of interest are the found seeds of *Polygonum convonvulus*. It is possible that the plant was used in the nutrition of the poultry as it is known that the fodder of *Polygonum convonvulus* is good for them. But *Polygonum convonvulus* is also used as (in the form of) groats or as admixture added to the bread flour. Such flour has a specific taste and the admixture results in a dark color of the bread.

The plant is found often in the bushes and in the grassed lands (fig.8).

DAUB ANALYSIS

As the quantity of the material was highly insufficient, daub analysis was done. It covered 1018 fragments sized from 3 cm to 5 cm. The correlation between the quantity of fragments and the found imprints is quite low. The big number of it refers to wild growing cereals and here and there were found also imprints of einkorn and emmer. In the first place were proved existence of chaff of *Bromus secalinus*, followed by barley.

The *Bromus secalinus* is a plant found quite often in grassy weed lands, grazed willingly by the cattle. Fixed in the daubs was also quite a good number of wild growing plants imprints such as of foxtail millet, different species

of *Vicia* and here and there of common millet. The material as a whole is quite restricted as quantity, the basic wheat cultures are slightly documented and the dominating plant is the bitter vetch.

SLATINO

Location and Physicogeographical Characteristic of the Region.

Dating, Excavation leader – St. Chohadjiev

The village is located in the downhill of the Northern Rila Mountain at about 380 m above sea level, near to Struma River.

It is dated from the Early Chalcolithic and it covers 5 building horizons. According to the archaeologist Chohadjiev (2006, pp.38-42) its dating shows **C** 14: 4650 – 4500 cal. B.C. The depth of each one of the layers is 1, 5 m. The area of the settlement is 2, 5 ha.

Contextual Analysis

The study materials present brown-reddish daub. They were collected from the II horizon of the central profile. All fragments are dated from the Chalcolithic period. On their surface are seen numerous imprints of cereal plants – glumes, stems, spikes and grains of cultivated cereals. A total 164 fragments was found with a total volume of 1439, 63 cm³.

Most of the imprints belong to the hulled wheat – einkorn and emmer. Except of wheat species fixed are also imprints of naked and hulled barley. Found are also numerous imprints of leaves of wheat, but their species could not be defined.

The charred plant remains show the presence of einkorn, emmer, bitter vetch, naked barley (Popova, 1995c). The data published by Marinova show the presence of emmer as dominating as well as a diversity of leguminous. Found were also some weeds species – *Bromus sp, Asperula arvensis, Polygonum convolvulus, Centaurea sp.* Most of them are characteristic for the winter sowings (Marinova, 2002). In the samples were found collected seeds of *Chenopodium sp.,* which probably were stored for periods of nutrition crises. The seeds of the *Chenopodium sp.* contain proteins and fat and supposedly they have had a definite usage.

TOPOLNITSA – Promahon

Location and Physicogeographical Characteristic of the Region.

Dating, Excavation leaders – H. Todorova, I. Vayssov

The site is located by the Bulgarian – Greek border about 2 km to the south of the village of Topolnitsa, Petrich district. It is situated 1 km to the west on the right bank of Struma River. The border divides it into two sectors – Bulgarian (sector Topolnitsa) and Greek (sector Promahon). The settlement was inhabited in the Late Neolithic period and the Early Chalcolithic period – about 5100 – 4700 B.C.

The presented there culture Topolnitsa – Acropothamos is an important linking unit of the Late Neolithic events in the South-Western Bulgaria to those in Northern Greece lands, in Macedonia and Thessalia (Koukouli-Chryssanthaki, et al., 2007).

CONTEXTUAL ANALYSIS

The materials are collected from different archaeological contexts and part of it is daub collected in the III-rd horizon of the central profile - dwelling No 3. The volume of the studied material is 195, 75 cm³.

The daub contains admixtures of straw in the form of glumes as well as several imprints of grains but they are not sufficiently distinctive. Small part of it is defined as einkorn. In that connection it could be supposed that hulled wheat was grown in the settlement but more exact characteristic of the species contents could not be done because of the lack of sufficient quantitative data.

In the site was found both charred and uncharred wood. The uncharred wood belongs to horizon II, house No 2. It is strongly mineralized. Another group of samples is collected from layer No 36 – from an wooden construction and 2 samples from layer No 37. A digital microscope was used for the definition of the mineralized wood fragments–targeting to define the anatomic specifics of the species of wood. The analysis of the material was compared to atlases and collections (Gregus, 1955; Gregus, 1959; Schoch, et all., 1988).

The results of the study show that oak fragments – *Quercus sp.* were predominant and only a small quantity of *Rosaceae* fragments was found. Three fragments are referred to pine tree species – *Pinaceae,* and two of them are of *Pinus sp.* All these fragments are with significantly damaged structure and their identification is quite difficult. In the study of these samples were studied also wooden tools–nails – one of them made of oak, the other of the wood of *Rosaceae* – plums, pear or a similar fruit tree. The wood in the samples of layer No 36 and layer No 37 is of oak.

The analysis of the charred wood provides useful information for the study of the natural environment in different archaeological periods. It shows that the usage of wooden material of man has been a single permanent process.

In that connection the anthracological analysis is of definite significance. But for achieving reliable results from palaeoecological point of view it is necessary the collection of sufficient quantity of charred wood from different archaeological structures to be done.

In the studied materials collected in the settlement of Topolnitsa we could confirm the dominating role of the oak tree as well as the presence of some other species of wood as the pine - most probably it is black pine, and of the broad-leaved – maple (*Acer campestris*) and *Rosaceae*. However, an entire picture of the palaeoenvironmental conditions is hard to be seen.

HOTNITSA

Location and Physicogeographical Characteristic of the Region.

Dating, Excavation leader – St. Chohadjiev

The tell settlement is situated at about 1 km north-east of the nowadays village of Hotnitsa, Veliko Tarnovo district, in the left bank of Bohot river. The diameter of the hill by its basis is about 110 m and its height is about 5 m. The initial excavations were carried out by N. Angelov at the end of the 50-ties of the XX century. The excavations were restarted in 2000 by a team from the University of Veliko Tarnovo lead by St. Chohadjiev. They covered an area of about 200 sq. m. in the northern part of the hill. Till recently are studied the II-nd and III-rd buiding horizons. Partly studied were several dwellings which had gone through fire and where in one of them a big quantity of wheat was found. Their dating is of the Late Chalcolithic period. Chohadjiev and Elenski (2002).

The archaeobotanical analyses are part of the interdisciplinary studies of the tell settlement of Hotnitsa, which has been undertaken systematically already for several years. The samples were collected at the time of the excavations. For which I express my gratitude to the senior researcher St. Chohadjiev who provided me with them and to the team of university students who joined in the activity.

CONTEXTUAL ANALYSIS

The material originates from different archaeological structures: from and from around ceramic vessels, roof constructions, utilities, ovens, dwellings floors – where mainly were studied are house No 9 from the I-st building horizon, house No 1; house No 10 and, from the sq.14 house No 4 and house No 6. All samples were manually collected. No flotation was done. The quantity of the collected material is really huge.

TAPHONOMY

4 samples are taken from house No 9. The following species of plants are established: einkorn - *Triticum monococcum* L., emmer – *Triticum turgidum L. subsp. dicoccum* (Schrank) Thell., naked barley - *Hordeum vulgare var. nudum.*

The samples taken from house No 10 contain about 100 grams clean naked barley. In house No 4 was studied a fragment of straw. The straw had burnt and it was mixed with compacted charred daubs. Very attentively small branches presenting spike stems were separated from it. The separation made it possible these stems to be defined as stems of einkorn.

In the same house were found big quantities of mixtures of einkorn with emmer. Emmer was dominating.

TAPHONOMY

The analysis of the material from house No 4 shows domination of the naked barley. The barley is very well preserved and it is with huge grains. It is 6-row barley. In some of the samples from house No 9 were found single grains which are also with characteristic features for the 6-row barley.

The quantity of the naked barley is huge - about 5 kg. The grains are cleaned from the glumes. Probably the case here is a big storage targeted either for sowing or for direct use.

The einkorn is the second species of plant defined in the studied house. In house No 9, judging on the basis of the

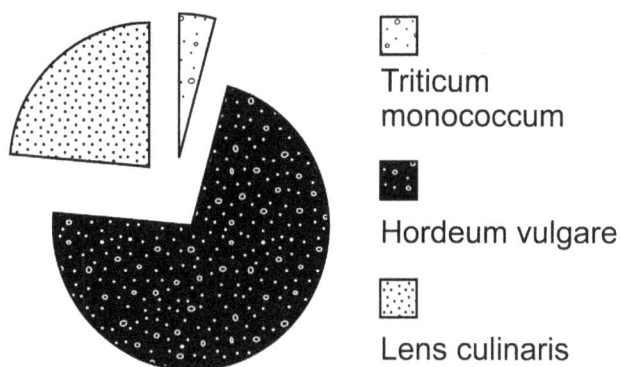

Fig. 9. Species composition of the cultural plants in dwelling 4 –Hotnitza.

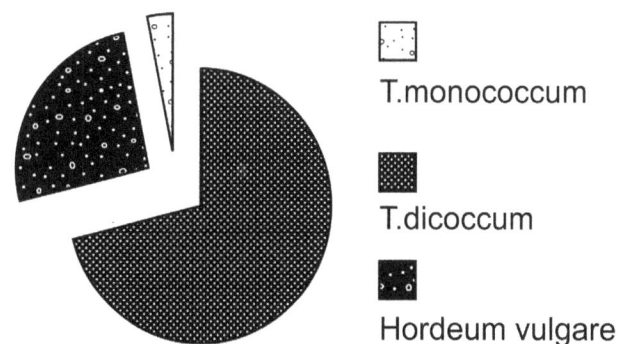

Fig. 10. Species composition of the cultural plants in dwelling 9 –Hotnitza.

Fig. 11. Chik –pea –Cicer artietinum – carbonized seed from Hotnitza.

quantity of the found material, the einkorn was also storaged. Ranked second, emmer is also presented together with the einkorn. The both species present a mixture dominated by the einkorn. The grains are very well preserved, cleaned from the glumes, there are no weeds. It is obvious that the mixing took place later and the mixture was ready for consumption.

The hulled wheat has quite strong stems. Their strength is due to the siliceous elements that turn them into unbreakable. The contents of these elements in the hulled wheat are higher than in the other species of wheat. In that connection they were very often used as construction material. As confirmation of it are the found fragments of einkorn

Fig. 12. Imprints of bracken fern.

stems, mixed with clay, which were part of the roof construction.

In general the impression is that there were big grain storages in these dwellings – both of wheat and barley. The materials from houses No 4 and No 9 refer to Class A.

By comparing the two houses some interesting differences could be outlined (see figs. 9-10). In house No 4 were stored only barley and bitter vetch when in house No 9 the stored goods were only einkorn and emmer.

Some authors say that the 6-row barleys in the past were sown in autumn. Additionally there are data that the biter vetch was sown among the barley plants. Then we should suppose that they were stored together after the collection of the yield. In confirmation of this hypothesis could be presented the data published by Marinova (2002, p. 178-179) who has also found stored barley, bitter vetch and wheat in a dwelling in the tell settlement Karanovo (Karanovo III).

In two houses (No 1 and No 6) was found chick-pea. In the territory of house No 1 the seeds of chick-peas were found among those of the barley and in house No 6, sq. 96-97 the chick-peas was found as admixture in storage of einkorn and emmer. Single grains of chick-peas were defined also in a ceramic vessel. The main content of the vessel was barley. Here and there were found also seeds of peas and lentils.

The plant chick-pea *(Cicer arietinum)* was not well known till recently by archaeobotanic data. It was documented from the Roman time in the settlement of Nikopolis ad Istrum . So it is of special interest for this period. Its presence in those houses as well as in the Chalcolithic layers of the tell settlement Junatsite provides evidence that chick-pea arrived in the territory of Bulgaria together with the whole Anatolian complex of cultural plants.

The findings of chick-pea among the pulses plants are of special interest as till recently it was known to be found only in the southern parts of the Balkan peninsula, and namely in Dimini and Otsaki (Kroll 1979, 1981).

GATHERING

In general we observed abundance of acorns and cornel-cherries in this settlement. Their quantity is quite big.

The cornel tree is easily accessible as it grows by the forest ends. Its habitat is opened to sun places and the bushes in the mixed oak forests. Its wide distribution in the low mountain ranges makes it an attractive plant both for its cherries and for its wood. The cornel tree is one of the most often found trees that mark all prehistoric periods in the territory of the country. Its ligneous fibers are of extreme strength. Except of the above mentioned fruits in the settlement were found also fragments of cherries and plums thus providing evidence for a certain collection in the region.

From the analyzed samples of interest are also the found imprints of bracken fern inside a pithos. Such finds happens for first time in the Bulgarian lands. The bracken fern grows in the bushes and in the forests. It is known that its leaves are used as litter/flooring in the cattle-sheds and that they improve the quality of the manure. Due to the specific aroma of the leaves and to their anti-decay qualities the leaves are used for coverage of fruits and vegetables. It is very possible that the bracken fern was used for conservation of some nutritious products.

Charred wood was found in almost all taken samples. The quantity of the fragments is different and their dimensions also vary. Established as presence are 3 species of trees: oak, maple and ash-tree. The oak dominates. Observed are traces of whittling on some oak and maple fragments.

As concerning the charred wood, the dominating species was the oak that was widely spread in the past as it has very good qualities for usage both in construction and in production of different items.

The presence of ash-tree in the region could be explained by the more humid conditions in the past as the plant grows well in cooler places and in soils with constant humidity. Its wood as that of the maple is very fit for the elaboration of small wood proceeding tools.

The tell settlement Hotnitsa is rich of archaeobotanical material. In the studied site the emmer is presented in big quantities. The wide distribution of emmer is due also to its adaptability to different ecological conditions. The barley keeps the second place in the studied materials. In conclusion it could be said that the presence of these species of plants provides results similar to the results of other studied settlements of these periods in the Balkan Peninsula – those of Opovo, Okolishte and others (Borojevich, 1988; Bittmann and Kucan, 2004; Valamoti, 2004).

BRONZE AGE

Early Bronze Age

GOLYAMA DETELINA – Burial mount II

Location and Physicogeographical Characteristic of the Region.

Dating, Excavation leaders – M. Kanchev, K. Leshtakov, B. Borisov

Burial mount II is a part of a mount necropolis studied initially by M. Kanchev (from the History Museum of Nova Zagora) by rescue excavations which had been undertaken in the lands of the nowadays village of Golyama Detelina in the area of the energy complex of Maritsa Iztok. The necropolis is dated from the Early Bronze Age and Late Bronze Age graves were also found in the mount (Leshtakov, et al., 1995).

The archaeobotanical analysis ascertained insignificant quantity of charred oak fragments.

DURANKULAK – Bronze Age

Location and Physicogeographical Characteristic of the Region

Dating, Excavation leaders – H. Todorova, Y. Boyadjiev

(The description and the dating of the site are provided with the Chalcolithic materials.

CONTEXTUAL ANALYSIS

In house No 3 A were found several species of wheat – einkorn, emmer as well as hulled and naked barley. Among them were found also several grains of bread/durum wheat – *Triticum aestivum/durum*, wheat - spelt – *Triticum spelta* and 1 grain of wild einkorn wheat –*Triticum boeoticum Boiss*. Found were also seeds of biter vetch.

The quantity of this mixture is insignificant. It should be a case of accidental mixing as result of splitting corn on the floor of the dwelling. The materials refer to Class C.

As elements of gathering appear the found fragments of cherries - *Prunus avium* Moench and *Prunus cerassus* L.

The seeds of elder are also in this period. These finds provide evidence that, together with growing cultural plants, the collection of all kinds of fruits in the vicinity of the settlement where those plants were found was a common practice.

MADRETS – Layer A – Early Bronze Age

Location and Physicogeographical Characteristic of the Region

Dating, Excavation leaders – K. Leshtakov, I. Panayotov

(The description and the dating of the site are provided with the Chalcolithic materials).

CONTEXTUAL ANALYSIS

In sq. No B 2 from floor level were found the following species of cereals: einkorn, hulled barley, common millet. Except that there was also found a substantial quantity of bitter vetch. The weed flora is presented by the following species: *Chenopodium album, Gallium spirum, Polygonum aviculare,* etc. It is known that parts of these plants – leaves and seeds are used as nutrition supplement. The seeds of the *Galium* are a good food for the poultry. The status of these plants is still not clear – if they were collected intentionally – but it is accepted as fact that they were used for consumption (Hubbard, 1980; Gones, 1992). The seeds of the goose-grass are a good food for the poultry.

The elements of collection are documented by the presence of stones of cornel tree fruit, plums, acorns, cherries and grapes.

TAPHONOMY

The character of the material provides evidence for different activities. It is manifested by the different mixtures of wheat grains with leguminous and wild growing plants. They could be referred to Class C.

In conclusion we could ascertain the presence in all samples of einkorn and hulled barley. All grains of cereals and leguminous are quite deformed and fragmented.

We should note that the bigger part of the presented weeds *(Chenopodium album, Gallium spirum, Polygonum aviculare Polygonum convonvulus, Rumex acetosa, Lithospermum arvense, Sanguisorba minor, Salvia officinalis, Convonvulus arvense, Veronica hederifolia)* are annuals plants and they characterize spring sowings - *Agrostemma githago, Bromus secalinus,* and that they are also very good indicators of well developed agriculture. On the other side

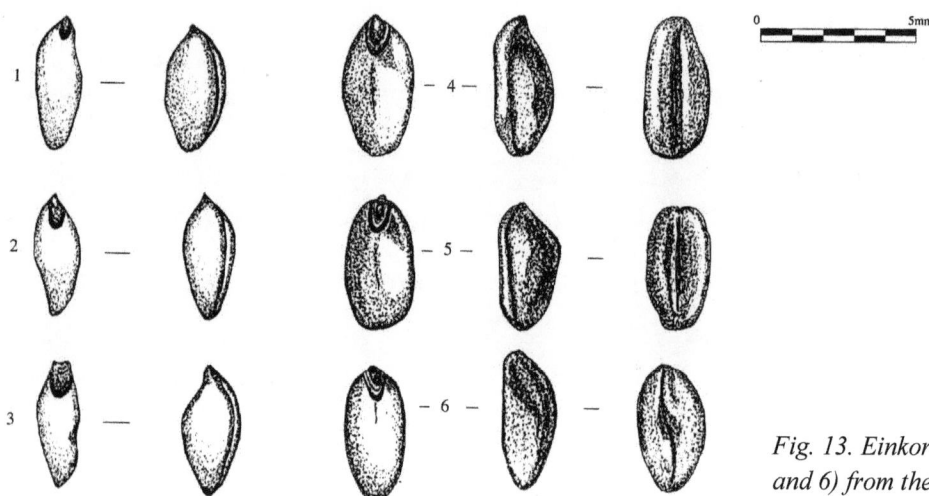

Fig. 13. Einkorn (1, 2 and 3) and emmer (4, 5 and 6) from the tell settlement Yunatsite.

hand different parts of most of those plants were put into use – the leaves, the seeds and the roots.

ANALYSIS OF THE CHARRED WOOD

In most of the studied samples was established the presence of oak wood.

Comparing the two layers in layer A could be seen an evidenced diminishing participation as percentage of the of the species. But that of the oak and the maple stays almost equal. To compare the studied material of the profile we took samples from a hearth. The samples are taken from different consequent depths by its direct dismantling as follows: by a layer with thickness of: 1, 46 cm; 1, 43 cm; 1, 3 cm. The following species of trees were established: oak, maple, ash-tree and hornbeam. The maple and the hornbeam were also found in the samples of the Chalcolithic layer. It proves that this vegetation was presented in the region during the both periods. These species are indicators for partly cut forests and they are often met by the river beds. As results the same species were established: *Quercus sp.*; *Acer sp.*

The number of fragments in the samples from layer A does not overcome 5 and a significantly higher number is established in layer B – about 15-20 in a sample.

It was established again that the oak was the most commonly used species of wood. It was found also in the samples of the other studied by us sites: Galabovo, Gledachevo, Dyadovo, etc. (Popova, 1995a, b). Accompanying species of trees are: *Acer sp.*, *Acer campestre*, *Ulmus sp.*, *Pomoideae*. They are typical representatives of the vegetation, growing by the river beds or on ruderal lands. The birch, the hazel tree (*Corylus avellana*) and the alder tree formed an association characteristic for the more opened and heliophilous spaces. The presence of species as pear, plum, blackberry and cornel trees, vogel beere *(Prunus sp., Sorbus sp., Cor-*

nus sp., Cornus mas, Sambucus ebulus, Prunus domestica ssp. institia, Prunus avium, Vitis sylvestris Gmell, *Quercus sp.)* and others document an active gathering.

YUNATSITE

Location and Physicogeographical Characteristic of the Region.

Dating, Excavation leaders – Ya. Boiyadjiev, Ya. Aslanis

The tell settlement is located 1,5 km to the south-west of the nowadays village of Yunatsite in the 4-5 m high terrace by the one of the branches of the former river bed of Topolnitsa River. The village is situated in the right terrace of the recent river bed of Topolnitsa, 2,25 m above sea level, about 8 km west from the town of Pazardjik. The soils around the hill are alluvial-delluvial. They are fertile and favorable for agriculture.

The tell settlement is oriented east-west with a maximum diameter of 110 m. Its average height is 12 m. Established are 4 cultural layers – remains of settlements with constructions from II-I century B.C., a Thracian settlement from VI-V B.C.; settlement from the Bronze Age, a settlement form the Chalcolithic. The excavations were carried out by R. Katincharov and N. Merpert (1967) and further on continued by a Bulgarian-Greek team – V. Matsanova, Y. Boyadjiev and St. Ignatova from Bulgarian site and Y. Aslanis of Greek site (2001-2008). The excavations still continued.

CONTEXTUAL ANALYSIS

The archaeobotanical material covers the layers from the Chalcolithic and the Bronze Age. Here are presented those from the Bronze Age as the materials from the Chalcolithic are unsatisfactory owing to circumstances beyond our control.

The archaeobotanical material is collected during several seasons and it presents huge quantities of charred seeds, grains and charred wood. The degree of conservation is different in the different samples as well as it also depends on the different context in which the sample was taken.

Studied are 81 samples from 15 horizons. The horizons from XVIII to X refer to the Early Bronze Age, IX and VIII – to the transitional period between the Early and Middle Bronze Age; VII till III – Middle Bronze Age. Most of the material was collected in the excavated squares of the corresponding horizons. In XIII horizon referred to the Early Bronze Age the material was collected from house No 22, 1 where 2 ovens and several pithoses were found. Obviously there was a big granary. The material taken from the XV horizon originates from several houses. In XVI horizon a huge grain store was found where the found charred wheat was measured to be 50 kg.

From the studied 15 building horizons of the tell settlement in 10 was found significant quantity of einkorn and only in 3 horizons were found single grains of it (XII, XIII, XVI building horizons). A pure sample of einkorn though was not found. It was almost always mixed with emmer. In all studied samples 2/3 of the grains are preserved and 1/3 are more or less broken.

In general some substantial declinations from the characteristic morphological features were not observed. The grains have the typical for their species form so they are easily definable. Their back side is asymmetrically hunch-backed and their front side is protruding. The average dimensions of the grains from the studied horizons vary insignificantly. It was established that there is no big difference in the dimensions of the einkorn from the Early and from the Middle Bronze Age in the settlement. The attention was given to the standard deviation in its (L) length and in its thickness (T). The length of the grain varies between 4, 00 mm to 6, 00 mm. The indicators for height vary between 1, 9 – 3, 7 mm and for the width between 1, 2 – 3, 2 mm. Differences in the grain dimensions in the transition period from Early to Middle Bronze Age are not observed. The average arithmetic L, B and T for the transition period does not show any aberration. The percentage correlation of the found einkorn grains in all samples to the rest of the wheat species is 33, 38%. Grains of wild einkorn are described by Chakalova and Sarbinska (1984) by the study of the tell settlement Kremenik. The description of the found and defined as *Triticum boeoticum* Boiss. grains by these authors coincides with that of the grains found in one of the samples of the tell settlement. The latter of the grains are quite narrower than those of the einkorn, their height is bigger and their germ is prolonged, their basis and most of all their tops are prolonged as well. Based also on the morphometric parameters of the measured 91

grains from the studied site, dated from the Early Bronze Age, it could be suggested that they belong to wild growing einkorn. Many times in the sowing of the cultivated einkorn the wild growing einkorn appears as mixture in the role of weed. Some of the found and studied grains have transitional features close to one or to the other species. The index T/B% of the both species is with near values and still the wild einkorn shows some higher values. The presence of emmer in the studied site was established in 13 horizons. In the most samples it is admixed with einkorn. As single grains emmer is found in several samples from IV, VII, X, XI and XIV horizons. The percentage contents of the emmer found in all studied samples is 35, 45% and as such it is the highest compared to all the other wheat species. Emmer imprints are established in daub taken from the X-th and XVII-th horizons.

In some of the Early Bronze horizons were found fragments of spikes with transitional features referred to *Triticum spelta* and *Triticum dicoccum*, as some of the grains come closer to *Triticum dicoccum*, and some others to the *spelta*. In some places in the site were found mixtures of glums of *Triticum spelta* and grains of *Triticum dicoccum* which most probably should be indicative that either both species were sown together or they were later collected and stored together in common grain storages. Some substantial declinations from the characteristic morphological features for the species are not observed. From their back side the grains are asymmetrically protruded, and their front side is flat. A small difference is established in the dimensions of the emmer in the transitional period from Early to Middle Bronze Age only concerning the length of the grain. Concerning their width and height so substantial differences are observed both in dimensions and as variations in the standard deviation in the separate horizons, when concerning their length significant differences are observed both in the dimensions of the grain and in the variation in the standard deviation. The length of the grains varies from 5, 6 mm to 7, 1 mm; the indicators for width vary from 2, 9 mm to 3, 4mm and those for the height vary from 2, 4 mm to 3, 0 mm.

In comparison with the other settlements from the Bronze Age in the territory of the country the following conclusions could be drawn: by comparing the data from the tell settlement by the town of Nova Zagora from VII and VIII horizons with those from Yunatsite tell settlement, we can see a significant increase in the dimensions of the grains (Popova, 1991c) The conclusions from the comparison of the results for the end of the Early Bronze Age, according to the newly established periodization from the Ziganska Mogila and Ezero (Dennel, 1978) come to show that the dimensions in length, height and width are higher.

Another species of wheat evidenced in the tell settlement is the spelt. A distinctive feature of the wheat species spelt is the way of falling apart/decomposition of the wheat-ears from the spike and the connected with it basis of the spike. By the falling apart/decomposition of the spelt spike after it is ripe, the separate spikelets stay connected with the brokening central axis of the spike and by it the basis of the spike is always much wider than that of the einkorn and of the emmer. The glumes of the spelt also differ from the above mentioned species of wheat – they are harder, by the upper front part of the grain there is an edge with rare but visible cog/pinnacle with sparse but visible nerves. If the species should be defined only on the basis of found grains, what should be taken into consideration is the following: spelt grains are much wider than those of the emmer and their top is always wider than the basis. The top of the grain is blunt, widely rounded and the basis is tapering/ sharp-pointed.

Judging from the archaeological finds the spelt in the territory of Bulgaria should have been quite restricted.

In the studied samples spelt was found in 4 samples taken from the XVI horizon. There were found grains and glumes which provide the basis they to be defined as spelt. The found glumes are typical for the species. The remains of the spike stem are grown into the inside of the spike, which shows that the spike belongs to *Triticum spelta*. The article of the spike axis along its whole length stays conglutinated with the internal part of the spike. The percentage correlation of the spelt is significantly lower – 3, 09% in comparison to the other wheat species.

One of the species found in the tell settlement of Yunatsite is barley. It was found in 8 building horizons. Found are both grains and fragments of spikes of 6-row barley. The barley was defined as hulled barley and in two of the horizons – XIV and XVI – as naked barley. As pure culture it was found in samples from XI, XV, XVI and XVII horizons. In the other horizons it was mixed with other cereals.

Big quantity of fragments of spikes were found in XIII, XIV, XV and XVI horizons and as compact grain mass in XIV, XV and XVI horizons. The bigger part of the samples is in good condition. The established percentage of presence for the barley is 31, 36% with which it ranks third in the findings – after the einkorn and the emmer.

In general the found grains do not distinguish from the characteristic morphological features for the barley species.

TAPHONOMY

The preservation of the grains in the different horizons is good and the degree of charring is almost one and the same. In some samples from the V, VII, VIII, IX, X, XIII and XV horizons was found different as quantity compact grain mass from einkorn, resulting either from fast charring or from presence of water that sticks the grains together. In some of them are observed significant deformations of the caryopsises.

One of the samples dated Early Bronze Age presents part of the material taken from granary. The quantity of the found barley there is about 50 kg. What makes special impression here is that a big part of it is in the form of compact mass of whole spikes. It is clear that in this case we have evidence of a significant storage and yield. Supposedly the barley in these contexts was intended for sowing. Moreover, from this context we achieve also valuable information about the way of collecting the spikes/harvest – and namely if the practice was a high or low harvest. In this case we have a high harvest – the stems were not cut but only the spikes were cut and collected. The material from the granary could be referred to Class A and the rest of the samples in its bigger part – to Class C.

LEGUMINOUS

Grass pea - *Lathyrus sativus* L.

The grass pea was found in two samples taken from two different horizons of the tell settlement – in horizon VII and in horizon XI. In one of the two samples it is without admixtures. And its contents is measured about 1, 5 kg. Its morphological characteristics provide basis it to be defined as *Lathyrus sativus* L. The sample taken from the VII horizon contains also admixture of peas. Both samples are dated from the Middle Bronze Age. In the sample of the XI horizon (Early Bronze Age) *Lathyrus sativus* L. is also dominating but also single grains of the following species were found there: einkorn, lentils and bitter vetch. They are in a very low percentage they and most probably they appeared there only accidentally.

The average values from the first horizon vary from 4, 5 mm to 5, 1 mm and those from the second horizon from 4, 2 mm to 4, 8 mm. Most probably the climatic conditions then were more favourable for its growing than today. The percentage presence of the *Lathyrus sativus* L. compared to that of the rest leguminous is 36, 36% which makes it the best presented leguminous plant among all other leguminous sown in the fields of the studied past.

Lentils – *Lens culinaris var. microperma* Medik.

Lentils were found in more significant quantities in XI horizon from the studied here horizons. Lentils grains in smaller quantities are documented also in several samples taken from the VII and XIII horizons. In the rest of the samples lentils always appear as admixture to the wheat or leguminous cultures. Among the leguminous for its presence the lentils keeps the second place with 31, 45%.

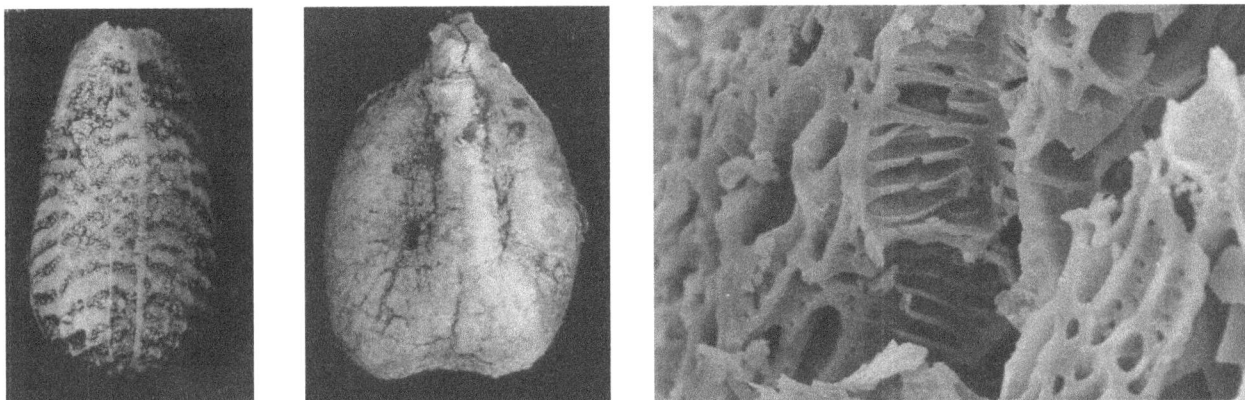

Fig. 14. Sambucus nigra (1), Vitis sp.(2) –carbonised fruits (3) Vitis sp. - charred wood.

Peas – *Pisum sativum* L.

Our study of peas is based only on a single sample taken from the VII horizon. The quantity of the peas is insignificant when compared to that of other leguminous – 18, 1 8%. It appears as admixture in the sample with the grass pea. The average dimensions of its diameter are: 3, 9 mm – 4, 0 mm.

TAPHONOMY

In general the grains of the peas are well conserved and the bigger part of them is of whole grains. The quantity of the broken ones is insignificant, what is logical as by high temperature the pea grains easily fall into parts.

On the basis of the morphometrical indicators the peas here are defined as suich from the cultural species - *Pisum sativum* L. Obviously the peas were also sown though on not so vast areas. Its dimensions are small, what is its characteristic feature also in other prehistoric settlements for the same period. That could be explained with the specific small size forms which were spread and were characteristic for the Balkan Peninsula.

Bitter Vetch –*Vicia ervillia* Willd.

The species bitter vetch was not found as a pure sample. Significantly higher quality was found in a sample taken from the X-th horizon. And, as single grains, bitter vetch was established in samples from the VIII, X and XI-th horizons. The general quantity of the found bitter vetch seeds presents 14% from all leguminous plants.

Fruit-bearing plants – cultural and wild growing.

In the studied material were not found big quantities of acorns, referring to III, IV and V horizons. Most of the acorns were broken. All of them were without cupolae. Envisaging the rich yiled and the huge quantities of stored

wheat, the acorns here were hardly added to the flour. So obviously they were used for the animal nutrition.

From the elements of collection the following were evidenced:

Sambucus nigra L. (elder tree)

Significant quantity of seeds of this plant was found in a sample taken from XIII horizon – Early Bronze Age. The seeds are without any admixture. The elder tree grows in the bushes and in the forests but also in the settlements.

The fruit of *Sambucus nigra* L. is a small suculent fruit. Its blossoms contain about 0, 025% essential oils, among them turpentine and paraffin oil. *Sambucus nigra* L. is one of the most popular plants of all those used in the traditional medicine. All its parts enter into use for treatment. The fruits are with abundance of vitamins A, C, C2 and are also used in the preparation of blue color for painting of cloth (fig.14.1).

Vitis viifera L. (vine)

Only a single seed of vine was found in this tell settlement. Still it was defined as the cultural species - *Vitis vinifera* L. with measured dimensions as follows: length – 4,8 mm and width – 3,4 mm. The vine seed was found in a sample taken from XIII horizon and it is dated from the Early Bronze Age (fig. 14.2-3).

WEEDS

By the study of the charred materials were found also some seeds of weeds, mostly from the families: *Asteraceae, Caryophyllaceae, Polygonaceae, Poaceae*. The most commonly found weeds are from: *Centaurea sp., Polygonum aviculare, Bromus secalinus.*

EARLY BRONZE AGE

YAZDATCH

Location and Physicogeographical Characteristics of the Region.

Dating, Excavations leaders – I. Panayotov, M. Hristov, R. Mikov

The site is situated in the place 'Kurutarla' near to the nowadays village of Yazdach, Chirpan municipality, the site No 16 in the planned highway 'Trakia' Lot 1. The studied lands there cover almost 13 dca.

Documented were over 50 negative archaeological structures. The most numerous structures refer to the Bronze Age: Early Bronze Age: I – 20; Early Bronze Age III: – 4; Early Bronze Age without opportunity for chronological definition – 18; Late Bronze Age: – 1; 5 structures refer to the Late Antiquity and 9 structures stay with undefined dating. Those from the Early Bronze Age in the most cases are middle depth pits with different dimensions, sunk in the soils and situated in the whole studied area. There are 4 studied pits from the Early Bronze Age III – No No 44, 45 and those in square B 10 and in square B 12/B 13. They are also spread over the entire area of the site. From the Late Bronze Age is studied only one pit No 70 A, which is cut by another later dug pit due to which its bigger southern part is destroyed.

According to Panayoptov, et al., (2005, pp.78-79) the site is dated from the Early Bronze Age. The excavations are carried out by the archaeologists K. Panayotova, R. Mikov and I. Panayotov in two consecutive years. The site presents a pit field.

The samples are extracted from those negative structures according to the established methodology for extraction of sediments from pit structures. Studied were samples taken from 56 pits. In most of them there were no contents of plant remains.

TAPHONOMY

The charred grains are small sized. They are deformed and with poor conservation. Such deformation is achieved by the depositing of materials and later due to the different degree of burning. Of significance is also the very process of burning - if it took place outside the pit or inside it, which is of significance for the conservation of the material. The following species of plants were witnessed: naked barley, rye, oats, and common millet. The dominating culture is the barley and ranked second is the rye. Often found is also the *Bromus secalinus*. The amount of the sediments, which have been studied by means of flotation,

and the quantity of samples were sufficient but as result were found not enough plant remains. The materials from the pits refer to Class A.

LATE BRONZE AGE

NEBET TEPE

Location and Physicogeographical Characteristic of the Region.

Dating, Excavation leader – A. Peykov

The settlement is located in the old part of the city of Plovdiv. It comprises 3 basic horizons which, according to the opinion of A. Peykov, are dated in the following way: horizon I – Early Bronze age culture Michalich, horizon II – Late Bronze Age and horizon III – Early Iron Age.

CONTEXTUAL ANALYSIS

The materials cover different archaeological contexts comprising the three horizons and the materials are collected in their biggest part from floor levels. There are found weed, naked wheat and barley, common millet and leguminous plants.

The achieved results from the measurement of the einkorn seeds provide evidence that there are not big differences in their dimensions depending on the horizons. Definite changes are observed only in the length of the einkorn seed – from 4, 6 mm to 6, 0 mm, while by the height the alteration is much smaller – from 2, 7 mm to 3, 0 mm. Comparing the results with data from other authors the following picture is drawn: the width and the height are bigger and the length is smaller.

The emmer is also documented also in the three horizons but it is in a smaller quantity – 5, 5% in the second horizon. The sample most probably presents a secondary product.

Of special interest are also finds of grains of naked common wheat *Triticum cf. aestivum* which could be traced in all the three horizons. Ascertained is a significant variation in the size of the grains. The minimum for its length is 4,6 mm and the maximum is 5, 7 mm. Smaller differences are found in the width – from 2, 9 to 3, 4 mm and in the thickness – from 2 ,6 mm to 3,1 mm. The grains are relatively short and with lightly rounded ends. The presence of this wheat there shows that it started to get its defined role in the composition of the cultivated cereals.

Triticum compactum – This compact wheat is evidenced as admixture of the common wheat. Its grains are shorter than those of the common wheat and its index L/B is lower than 1, 5 which is characteristic for the compact wheat. Another naked wheat is the durum wheat (evidenced in II horizon). The documentation of these three species of wheat is of a special importance as they mark the beginning of the Bronze Age.

The barley grains belong to the species of the naked barley.

TAPHONOMY

The einkorn is traced in the three horizons and it ranks first by its quantity. In the I-st horizon (Maritsa culture) it is found in a mixture with different cereals and pulses. Its participation as percentage is 34, 78%. The sample presents a wheat mixture, collected from a floor structure. The quantity of the grains is insignificant and obviously they have fallen there accidentally as result of different economic activities. The materials refer to Class C.

In a vessel in the first horizon was discovered common millet. Several grains of oats were also found there. The oats grains have obviously fallen there also by accident. The grains are flat. Their glumes are completely burnt. The basis of the grain is pointed (cuspidal) and the top (pinnacle) is obtusely rotund. Unfortunately the glumes are missing so it is difficult to define if they belong to the cultivated oats or to the wild growing species as only on the basis of the presence of the so-called 'thickening' – horseshoe in the basis there could be judged for the belonging of the oats to the wild growing species. The materials in the vessel refer to Class A.

The presence of 8 grape grains referring to the layers of the Early and Late Bronze Age could be added to the elements of collection. The typical indicators for the distinction of the wild growing grape from the cultivated grape is the index 'length to width' – L/B. In the cultural species the length of the seeds is almost twice as big as the width and in the wild growing ones the value is from 1, 2 – 1, 4, i.e. they have more rounded form.

Based on these indicators it could be judged by the index that the seeds found by Nebet Tepe belong to the cultural species, and still their lesser relatively small length brings them near to the wild growing. That is why we assume that the found charred seeds are of hybrid origin as their index is 1, 5 : 1, 6 which is often found by hybrid forms between the wild and the small-sized seed cultural forms. As for the contextual analysis of the samples we could not obtain more precise information owing to circumstances beyond our control.

KAMENSKA CHUKA

Location and Physicogeographical Characteristic of

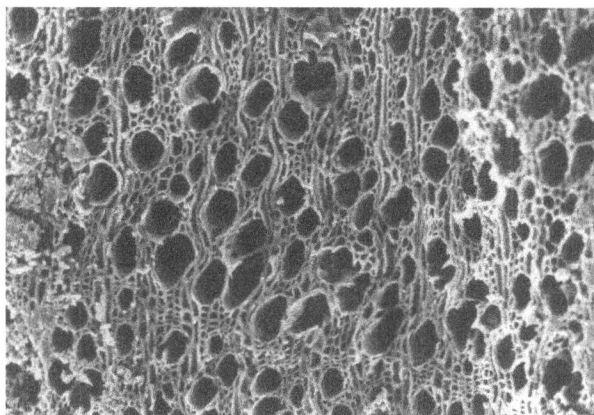

Fig. 15. Pomoideae - Charred wood from Kamenska chuka.

the Region.

Dating, Excavation leaders – M. Stefanovich. I. Kulov

The settlement is located near to the town of Blagoevgrad. It has implemented protection function and it is located 404 m above sea level. From geological point of view its basis consists of alluvium soil surrounded by coluvium slopes (and metamorphic hills and rocks (Stefanovich and Bankoff, 1999). This settlement is dated from the Late Bronze Age.

CONTEXTUAL ANALYSIS

The materials were collected in the course of 4 seasons. They comprise sediments of different archaeological contexts – mostly gathered from dwelling floors and from the contents of pithos (archeological units – unit No 2, 3, 5, 6, 12, 13, 25, 54 etc.; pithos in sq. 12 – north-western corner of the square) and they refer to the Late Bronze Age.

69 samples were studied altogether. The samples are extracted through flotation, assisted by a special flotation machine. After the separation of the individual fractions and their dehumidification it was found that the quantity of the plant remains such as seeds and grains is insufficient, while the presence of charred wood fragments is quite representative.

The result showed that in 15 from all 69 samples there were contents of plant remains.

TAPHONOMY

The quantity of the grains from the house floors and from the pithos is small but the dominant taxons are definitely as follows: *Triticum cf. boeoticum* – wild growing einko-rn; *Triticum monococcum* L.- einkorn; *Triticum dicoccum* Schrank.- emmer; *Hordeum vulgare var. nudum* L.- naked barley; *Panicum miliaceum* L.- common millet; *Vicia er-villia* Willd.- bitter vetch; *Lens culinaris* Medik.- lentils.

Einkorn is the dominant species – *Triticum monococcum*. Species as lentils and common millet were found in insig-nificant quantities. The representatives of the weedy her-baceous plants are documented by the following species: *Chenopodium album; Polygonum aviculare; Gallium sp.; Rumex acetosella, Centaurea sp.*. The materials of this con-text – pithos in square 12 – could be referred to Class A, while those from the house floors refer to Class C – the material is result of several different events connected with burning and/or different activities. The plant remains there have only slight connection with the context where they were found.

In most cases the grains are with poor preservation and quite often they are broken. Big part of them is in mixtures and they are collected from different archaeological struc-tures, i.e. they appear as a secondary product of different economic activities. Single grains of different species were also found, which provides evidence of their accidental presence.

The quality of the material proves that it originates from several different economic activities so the materials refer to Class C.

In contrast with it in another pithos – pithos No 3 in sq. 13/16 were found about 100 g blackberry seeds - *Rubus sp.* as well as such of blue elderberry– *Sambucus nigra*. The question here is what was their purpose. The first pos-sible suggestion is that blackberries were collected as fruit and later due to a process of carbonization only their seeds were preserved and that is because the succulent parts of the fruit could be hardly preserved by carbonization. And the second possibility is that they were gathered by an ani-mal as its storage. These plant remains refer to Class A.

The collection is documented by the presence of cornel tree – *Cornus mas;* elder – *Sambucus nigra*; elder tree-*Sambucus ebulus,* cultivated grapes – *Vitis vinifera,* which prove the usage of these plants for nutrition.

The grape seeds are determined on the basis of their mor-phometric features which bring them near to the cultural sorts of grapes.

CHARRED WOOD ANALYSIS

Charred wood was found in 30 of the samples. 618 wood-en fragments were defined. Ascertained are 16 taxons. The oak is the most often found species, accompanied by *Pinus*

and *Pinus syvestris,* juniper and pine spruce (*Abies alba).* Plant species as elm-tree, ash-tree and some representa-tives of *Rosaceae* are less often presented (fig. 15).

Most of the charcoal fragments were collected around the ceramic concentrations. The data show increased us-age of oak wood. The other species were lesser used. The different percentage participation of the species could be explained with the ways the materials were collected. The oak is found at lower heights, it has valuable constructive features and it could be burnt for domestic reasons while most of the other above mentioned pine species, for ex-ample the *Pinus syvestris* are found in more remote places and by higher altitude.

POLSKI GRADETS

Location and Physicogeographical Characteristic of the Region.

Dating, Excavation leader - K. Nikov

According to the excavation leader K. Nikov the studied material refers to the Late Bronze Age and to the Iron Age.

CONTEXTUAL ANALYSIS

10 pits have been studied. In pits No 43, 48, 50, 64 and 65 the following species of plants were found: einkorn, common/durum wheat and barley. The material is insuf-ficient for definitive conclusions. It could be only noted the appearance of the bread/durum wheat whose spread is connected with later periods.

The analysis of daub imprints found in the pits No 46, 47, 49, 53, 55 and 58 shows analogy in regard of the species composition.

ADATA

Location and Physicogeographical Characteristic of the Region.

Dating, Excavation leaders – P. Georgieva, B. Borisla-vov

I was provided with the archaeobotanical materials from this site thanks to the kind collaboration of the archaeolo-gist Dr. B. Borislavov, for what I cordially thank him.

Supposed dating of the site – from the Bronze Age.

CONTEXTUAL ANALYSIS

The samples include daub fragments collected by the surface of the site from an area of about 11 squares during the excavations. Except them 12 samples from pit No 1, extracted by flotation, which present soil deposits.

The samples were taken consequently by 5 cm in depth of the pit, comprising an interval of 1, 98 – 3, 02 m. Part of the content of pit No 3 in sq. H 12, pit No 4 in sq. J 4 and from sq. G 10 – from a ceramic construction – were also subjects of study, being extracted through flotation. The found seeds and grains are representatives of cereals, leguminous, fruit and weed species. In some of the samples from the pits was found also charred wood that was also defined.

ANALYSIS OF THE CHARRED REMAINS FROM PIT NO 1

The quantity of the flotation samples from the pit was satisfactory but the material revealed a quite poor picture. Single grains of barley and lentils were ascertained. In the sample taken from the depth of 1,99 m except barley was also found a cornel stone and a seed of *Cardius natans* - typical weed for the cereals that often grows by the roads. From the same pit were also extracted separate fragments of charred wood. In 10 of all samples is identified oak tree – *Quercus sp.* At 2, 00 m depth are found 16 fragments of *Carpinus cf. betulus.* The presence of pine trees is documented only through 1 fragment of *Pinus cf. sylvestris.* The biggest quantity of oak is found at the depth of 1, 99 m – 23 fragments. Also from this context were found two stones – of plum and cherry. The materials refer to Class A.

In the other studied contexts: from a ceramic construction in sq. G 10 at the depth of 2, 91 – 3, 00 m and 3, 02 – 3, 10 m there was also oak as well as in the contents of a cup found in the pit.

In sq. J 14 were ascertained charred integuments of pine cone and in sq. E 7 below the Chalcolithic layer, and under a heart were found charcoal and fragments of piths, hazelnuts. They represent two pairs of two halves but with different form and size. The first ones are rounded when the second pair of the halves is prolonged. According to their size here are presented two species: *Corylus avellana var. sylvestris* with small prolonged or almost egg-shaped fruit and *Corylus avellana var. grandis,* a species rarely found in nature but it is cultivated. It has big rounded fruit with thin cockleshell. The materials from this square refer to Class C.

Representatives of weed flora were documented with two species in drilling No 2 – *Rumex sanguineus* – 11 and *Chenopodium album* – 2.

ANALYSIS OF THE DAUBS

101 imprints were found over 471 daub fragments. The quantity of fragments in the separated squares is different. With higher concentration of imprints are sq. E 6 (71) and sq. I 8 (70). There were found 6 cultural plants species as common millet, barley, rye, einkorn and emmer, bread/durum wheat. In the sample from sq. E 6 the biggest quantity is of common millet – 26 imprints, followed by barley – 8. In sq. I 8, though with almost equal quantity of fragments, the imprints are less in number. There are found single grains of common millet, emmer, barley. That could be explained with the availability of less organic waste that was added in the process of preparation of the clay dough. In the other squares the species content is analogical and the quantity is insignificant.

ANALYSIS OF THE CHARRED WOOD

The data show that in the vicinity of the region the oak was frequently the most collected. The second preferred species is the hornbeam/yoke-elm – but it is documented only in a single sample. In a single sample were found also several fragments of willow and beech.

The studied material provides opportunity to be drawn the following conclusions: the data of the daub study show that the most often component is of common millet. Considering its ecological features as well as the location of the site and its altitude, the common millet was probably one of the basic sown cultures – because it is draught resistant, it exists by different environmental conditions and provides rich yield. The preparation of gruels of common millet as well as production of common millet bread is easy and obviously it was practiced quite often.

The found barley grains belong to the species of the naked barley. Barley is also a plant unpretentious to different climatic conditions. Finding rye is logical in this region and it should have been even in higher quantity as growing this plant requires colder climatic conditions.

The found species give opportunity to make some presumptions that these cultural plants were preferred in respect of their nature and the climatic conditions of the region.

From the weed flora were documented 3 species – *Cardius natans, Chenopodium album* и *Rumex sanguineus.*

DOSSITEVO

Location and Physicogeographical Characteristic of the Region.

Dating, Excavation leaders: K. Leshtakov, B. Borislavov

The site is located in the lands of the nowadays village Dossiteevo, Harmanli municipality. It is situated on a low but isolated hill in the valley of Balachka River (the Luda River), about 6 km from the Maritsa River.

The archaeoloigical site "Semera/Semercheto" is a Thracian cult complex of the kind 'rock sanctuary', used without interruption from the Late Bronze Age to the Hellenic time (Leshtakov, 2002).

The archaeobotanical materials were provided to me again thanks to the kind cooperation of the archaeologists K. Leshtakov and B. Borisslavov.

CONTEXTUAL ANALYSIS

The analysis comprises 8 samples taken from different archaeological structures from "Semercheto" site – from the profile South, layer No 12, from a stone concentration, from a vessel content, etc.

As a result of the analysis plant remains were found only in 4 samples. Studied was also the content of a vessel from drilling III, sector No 2.

TAPHONOMY

In there were found 2 common millet seeds. In another samples also from the same trench– profile South, layer No 12, was found a fragment of fruit stone – most probably of cherry/morello cherry. Unfortunately the typical morphometric features are missing and it puts obstacles to its definition. Several fragments of oak were found in trench V by a concentration in the Southwestern part (level II).

DYADOVO

Location and Physicogeographical Characteristic of the Region.

Dating, Excavation leader – D. Gergova

The tell settlement is located in the southern end of the village of Dyadovo, about 8 km to the south of the town of Nova Zagora, near to the highway crossing the country by its axis North-South. The tell settlement appeared at the end of the 7th and the beginning of the 6th millennium B.C. at the time of the Early Neolithic. The tell settlement is located on a small hill by the bank of the Blatnitsa River, near to two springs located in its north periphery where now there is swampy land. It has irregular oval shape with dimensions of its basis 220 m East-West and 140 m North-South. Its surface in its upper side covers about 8 dca. The depth of the cultural layers in its northern half is about 18

m, and due to the natural slope of the terrain in its basis, the cultural layers depth in its southern part is about 14 m.

The tell settlement of Dyadovo village is one of the biggest in Europe. It has been studied extensively over 30 years by Bulgarian, Dutch and Japanese archaeologists and that turned it into one of the most important sites. The studied materials refer to the Early Bronze Age.

The archaeobotanical studies in the tell settlement Dyadovo were run systematically in the last 10 years as much as it was possible from financial point of view when running the excavations.

CONTEXTUAL ANALYSIS

The samples are collected from different archaeological contexts. Some of them are from the central profile, mainly from sq. R 22. There was done stratigraphic antracological analysis. Samples were taken in every 10-20 cm.

Other part of the samples originates from IV and V building horizons. From IV horizon were studied samples from house No 1. Samples are collected also from the floor level of sq. M 29 (figs. 12-13).

TAPHONOMY

In most cases the grains do not have good conservation. Here were found naked barley *(Hordeum vulgare var. nudum)* and hulled barley *(Hordeum vulgare var. vulgare)*. In addition to them einkorn *(Triticum monococcum)*, emmer *(Triticum dicoccum)*, lentils *(Lens culinaris var. microsperma)*, and bitter vetch *(Vicia ervilia)* were found.

The ascertained seeds of lentils *(Lens culinaris var. microsperma)*, bitter vetch *(Vicia ervilia)*, einkorn *(Triticum monococcum)*, and emmer *(Triticum dicoccum)* refer to V horizon – in sq. O 19 from the contents of a pithos and they belong to Class A.

In the same horizon in house No 2 was found a small quantity of cereals consisting of einkorn *(Triticum monococcum)* and emmer *(Triticum dicoccum)*. Of special interest are the grains of emmer which are preserved in pairs together with their glumes. It shows that this mixture had been not yet cleared yet and was not ready for direct use, or it had been kept for sowing – thus, the material refers to Class C.

From biometrical point of view the analysis shows certain differences concerning the size of the einkorn seeds in horizon IV. There they are smaller than those from V horizon. Regarding the emmer grains, we observed a decrease in their length and increase in their height and width. Prob-

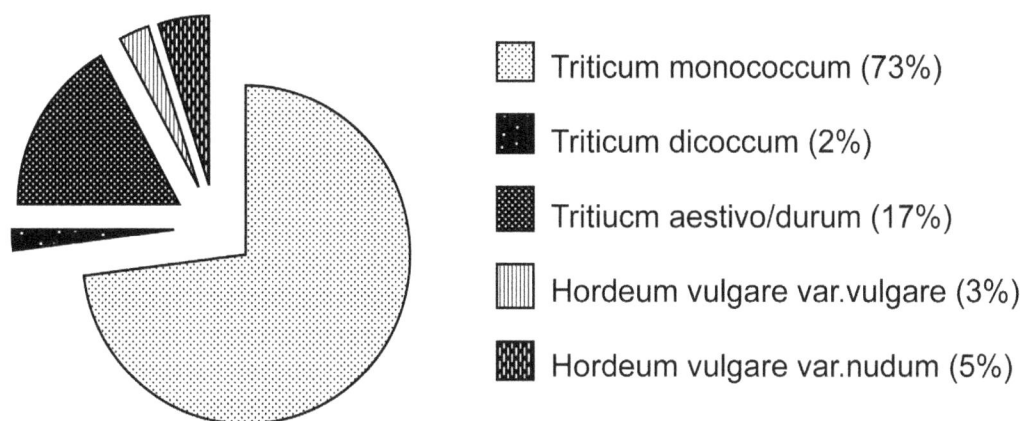

Fig. 16. Quantitative distribution of the charred plants remains in sq. M26 –tell settlement Djadovo.

Triticum monococcum (73%)

Triticum dicoccum (2%)

Tritiucm aestivo/durum (17%)

Hordeum vulgare var.vulgare (3%)

Hordeum vulgare var.nudum (5%)

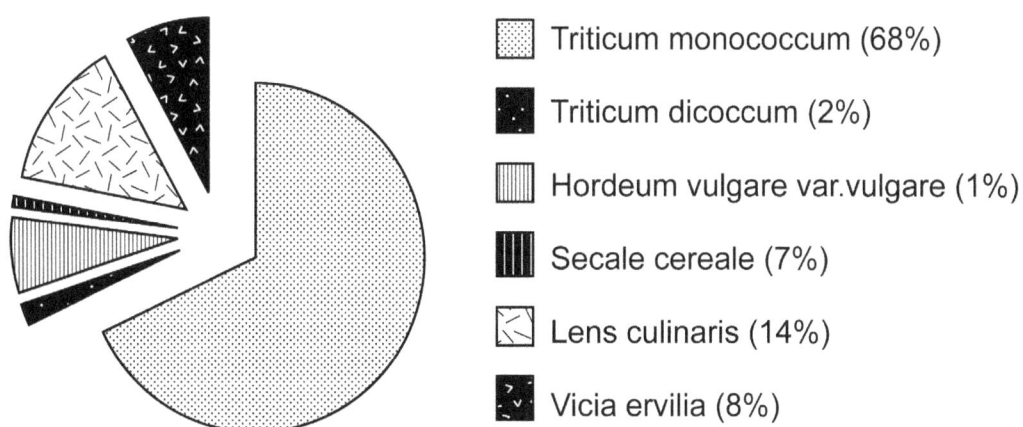

Fig. 17. Quantitative distribution of the charred plants remains from a padding an oven in sq. O 6 in the tell settlement Djadovo.

Triticum monococcum (68%)

Triticum dicoccum (2%)

Hordeum vulgare var.vulgare (1%)

Secale cereale (7%)

Lens culinaris (14%)

Vicia ervilia (8%)

ably it was due to a local population or it could be related to some other factors.

Small quantities of rye and oats were also found. Of special interest here is the appearance of the rye *(Secale cereale)* in the settlement. It deserves our attention as it is rarely met in the earlier periods.

The rye and the oats belong to the so-called 'secondary cultures' as their spread is connected to that of other wheat species. Rye moved north together with the bread/durum wheat. In this migration to more mountainous and colder regions the rye got separated as an independent culture also because it was more cold-resistant so it was used for single sowing.

Another cereal found in the samples, though, in an insignificant quantity, is the spelt – *Triticum spelta.* This species is rare to be found in the archaeological sites of this period, and only in minimal quantities: Neolithic –Durankulak, Nova Zagora (Popova, 1995c); Karanovo (Marinova,

2002b) etc.

The spelt does not require a special sort of soil. It is resistant to fungus troubles and to the damaging of its stems by birds. The spelt grains provide high quality flour. The grain is glass-like and very easily dehumidated after gathering of the yield, which is especially important for its growing in regions with humid climate.

Except in these archaeological contexts charred grains were found also in the samples taken from the central profile. As percentage the grains are minimally presented, here and there are found single grains of einkorn (*Triticum monococcum),* emmer (*Triticum dicoccum)* and barley (*Hordeum vulgare*).

ANALYSIS OF THE CHARRED WOOD

Charred wood was found in almost all archaeological structures and most often it belonged to oak – *Quercus sp.*

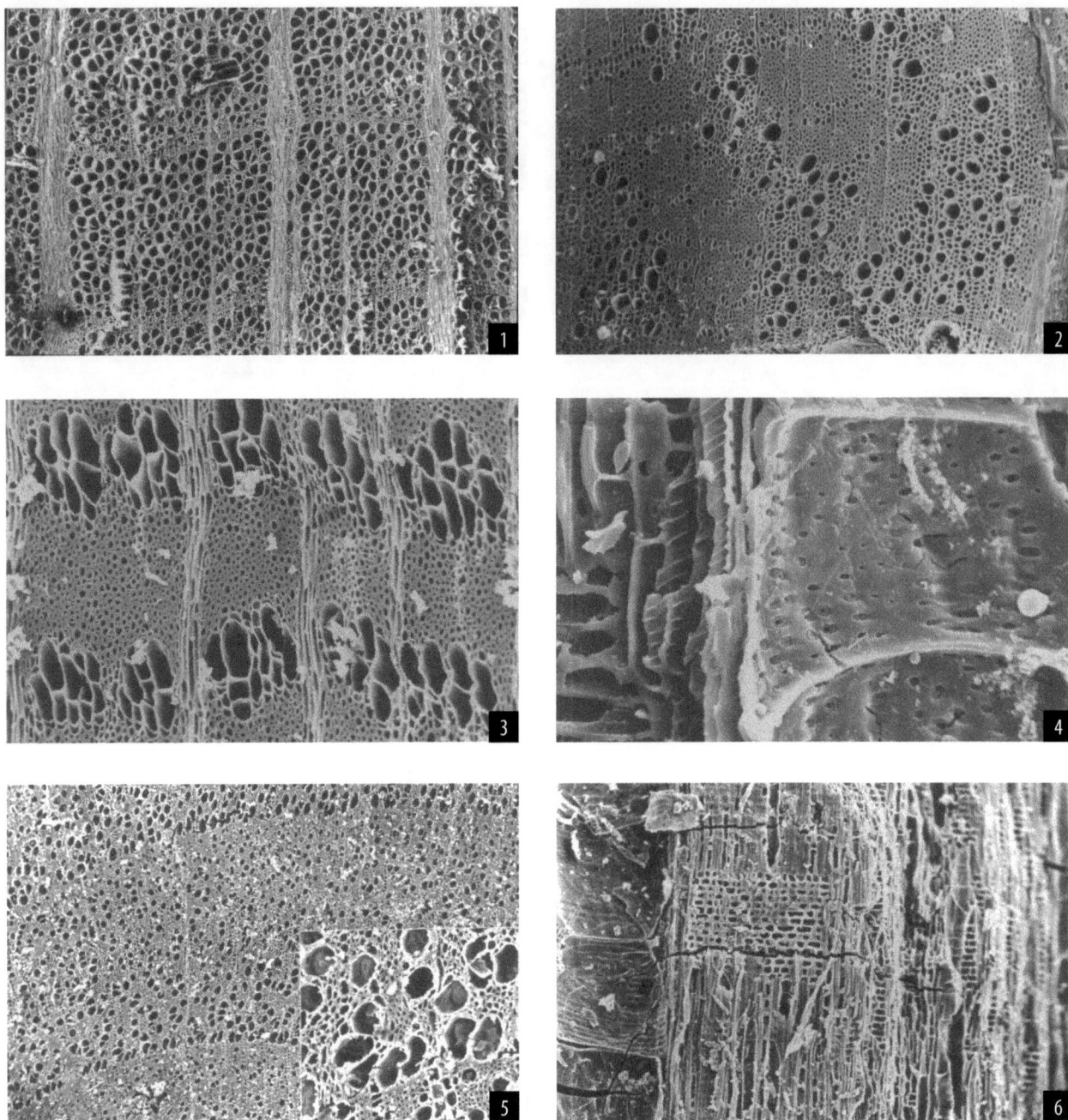

Fig. 18. Charcoal wood of: 1.Fagus sp., 2.Quercus sp., 3. Ulmus sp., 4.Ulmus sp.,5. Ostrya sp., and 6. Quercus sp.

The data from anthracological analysis of the central profile provide richer information about the usage of wooden material in the region. Found are the following species: beech (*Fagus sp.),* oak (*Quercus sp.),* hornbeam/yoke-elm (*Carpinus sp.*), elm-tree (*Ulmus sp.*), alder tree (*Alnus sp.*), fruit bearing trees (*Pomoideae*) (fig. 18).

Best presented is the oak tree – with 162 fragments, followed by the elm-tree and the hornbeam/yoke-elm. The oak is principally the most widely spread tree species here and it has been found in almost all studied archaeological sites till now. Its wood is a very good fuel as well as it is valuable

construction material. The features of the other tree species were given here above so I would not present them again.

From the above mentioned data and on the basis of the data of Bojilova, Chakalova (1980), which are easily correlated, some suggestion about the palaeo-environment could be done. It is obvious that dominant species were the oak, the elm-tree and the hornbeam/yoke-tree. According to the authors the later was in direct vicinity and, most probably it was an eastern species of hornbeam/yoke-tree. The beech as well has covered a substantial percentage in the samples.

Contextual and Taphonomic Analysis of the Studied Settlements

As a result the following picture could be presented:

The most commonly spread species were the different species of the oak. At highest altitude beech was collected. Alongside the river beds and in the more humid places the elm-tree and the alder-tree were the dominating species and consequently they were undoubtedly easier for collection. The found plant species in the archaeobotanical remains: oak – *Quercus sp.;* elm-tree – *Ulmus sp.* and hornbeam/ yoke tree - *Carpinus sp.* participate in the contents of the deciduous/broad-leaved forests.

Gathering

As elements of the collection were found seeds of elder *(Sambucus ebulus)*, cornel-tree *(Cornus mas)*, fragments of plum stones *(Prunus.sp)* and cherry *(Cerasum)* as well as of grapes *(Vitis vinifera ssp. sylvestris).*

Sambucus ebulus – the elder is found growing in the warmer places. The plant develops well in fertile humus soils and it grows mainly in cleared parts of the forests. It could be found also in dumping grounds/dunghills and alongside rivers.

MIHALITCH – Haskovo district, Svilengrad municipality

Location and Physicogeographical Characteristic of the Region.

Dating, Excavation leader: M. Stefanova

The site is situated in South Sakar, near the present-day village of Mihalich, Haskovo district, and it is surrounded by gullies to the south, east and west. The site was excavated first in 1945 by archaeologist V. Mikov. The excavations were renewed between 1998 and 2003 by M. Stefanova (the National Museum of History). The site is well fortified by a stone wall and the excavation revealed areas of domestic complexes with four levels dated from the Early Bronze Age 2 and 3 stages Mihalich and Sv. Kirilovo (Stefanova M., 2000).

The materials were recovered from the site during process of studying by (myself) the archaeologist M. Stefanova. Due to the limited excavations time as much as possible material was collected from the places during the archaeological work.

Contextual Analysis

The samples are mainly from daubs found in the following squares: sq. G 3, sq. F 2 and F 4 from Trench III. 26 samples were studied.

The analysis of the daubs showed predominant presence of einkorn, followed by barley. Single imprints of emmer and common millet were also documented. Regarding the charred remains several grains of einkorn were found in sq. G 3.

TATUL

Location and Physicogeographical Characteristic of the Region.

Dating, Excavation leaders – N. Ovcharov, D. Kodjamanova, Z. Dimitrov, K. Leshtakov

The rock sanctuary complex is located in close vicinity of the living quarter "Vejanitsa" of the village of Tatul, Momchilgrad municipality. The first excavations were done by I. Balkanski in the 70-ies of XX century. In 2004 the studies were restarted, carried out by the archaeologists N. Ovcharov and D. Kodjamanova. The earliest trace of human impact is from the period of the late Chalcolithic. After a certain hiatus the site was inhabited again in the Late Bronze Age, Early Iron Age, the Hellenic time, the Roman time, and the Middle Ages.

The analyzed materials originated mostly from the Late Bronze layer and they are related with its structures (Ovcharov, et al. 2008).

Contextual Analysis

All samples are obtained by means of flotation, and after that they were dehumidated and sorted.

The samples from Tatul site originated from 12 ritual hearths. After the analysis the following picture was revealed. 12 plant species were found all together. 4 of them are cereals, 2 leguminous, 2 weed, 2 fruit and charred wood.

The dominating presence is that of the einkorn (*Triticum monococcum*). Second is ranked the bread/durum wheat (*Ttitiucm aestivo/ durum*).

The leguminous are presented only with the lentils (*Lens culinaris var. microsperma*) and the grass pea (*Lathyrus sativus*). From the cereals except those two species are found also barley (*Hordeum vulgare var. nudum*) and emmer (*Triticum dicoccum*).

Taphonomy

The quantity of the found grains is very small and it could be explained with the high temperature of burning and with the connected with it ritual practices on the hearths.

The materials refer to Class A.

ANALYSIS OF THE CHARRED WOOD

Charred wood is documented mostly from the heartsh. The samples present only oak *(Quercus sp.)*. The oak wood is the most commonly found for several reasons.

The species of family *Quercus* are found in a low altitude and that makes them accessible for collection;

The wood material of *Quercus* is very good for burning as it reaches high temperature for short time;

It is very good construction material.

GATHERING

In the samples from several places were found fragments of acorns and cornel tree stones *(Cornus mas)*.

The cornel tree is one of the most often found plants in whole prehistoric periods in the territory of Bulgaria. It grows in the bushes, in the forests and in the rocky places of the country till about 1 300 m altitude.

Of special interest is the fruit of the stone pine *(Pinus pinea)*, found in one of the studied ritual hearths. The stone pine is a typical Mediterranean plant whose habitat is from Syria to Portugal.

The collection of cones for seeds seems to have been quite common in the Near East as well as in Cyprus and Greece.

The species was widely spread in the Near East already by the beginning of the Holocene. Especially in Italy it was used for culinary, decorative and even ritual designations. Many authors as Plinius, Collumela, Paladius report on the consumption of its seeds.

The stone pine grows by warm and dry conditions. It is more difficult for it to get accustomed for living to the north, due to which it is logical to have been brought to the interior of the Peninsula from the Mediterranean areas. The stone pine was burnt into hearths in ritual activities connected with the fertility. That was confirmed by the context of the found remains in Bulgaria – in Sozopol, Kabile, the Gyaurskata Mogila (near to the town of Carnobat). They were put into ritual hearths. Obviously this practice was widely spread and it was used because of the availability of ethereal and odoriferous oils.

From the weed flora were ascertained the following species: *Polygonum convonvulus; Galium apparine.*

Chapter 5

RECONSTRUCTION OF THE AGRICULTURE
AND THE VEGETATION IN THE STUDIED REGIONS

1. SOUTH-WESTERN REGION

Physicogeographical characteristic of the region

The South-Western region of the territory of Bulgaria was poorly studied from the archaeobotanical point of view until recently comparable to some other regions of the country. Over the last years the archeological excavations in this region increased. Most of the studied sites are located alongside the main stream and the feeders of the river of Struma.

The river of Struma is one of the biggest and most important rivers in Bulgaria. For its length of 290 km – measured from its sources till the state frontier with Greece - it is the 4th longest river in the country after the rivers Iskar, Tundja and Maritza. As for the size of its water catchment of 10 797 sq. km. – it stands second after the Maritza river. The Rivers Struma and Mesta are immediately connected with the sphere of influence of the Mediterranean zone – a fact that is essential for the structure of the water balance. This region is distinctive with the highest rainfalls compared to the other Bulgarian regions.

The most common in the region are the maroon forest soils and the most significant characteristics of the soil tegument here is the heavy erosion. The alluvium and delluvium soils alongside the river beds are shallow (Galabov, 1982).

The vegetation in the Struma valley westward of the town of Simitly in the Tundja valley south of Belitza, Slavjanka and the East Rhodopes is described as one with superiority of the Mediterranean vegetable life. The characteristic plant species for the region are: *Quercus coccifera, Pistacia sp., dendriform* and red juniper, *Jasminum offiunali*. A considerable place in the structure of the vegetation has the: *Quercus frainetto, Carpinus betulus, Ostrya carpinifolia, Ulmus campestre*. Many southern vegetation species are also represented there as: the red juniper, *Pistacia sp.*, the wild jasmine, the *Coronilla emerus*, the *Artemisia campestre* etc.

CHRONOLOGY

According to L. Pernicheva (1999) the development of the prehistoric cultures in south-western Bulgaria, and particularly along the Struma valley, is a key issue in Balkan prehistory. The direct territorial connection of this region with the northern Aegean coast, and from there with Anatolia, determined the specific dynamics of its development through all prehistoric periods (Pernicheva, 1999).

By the periodization of the pre-historic culture in the valley of Struma River the Neolithic period covers the appearance of the first ceramic till the appearance of vessels with graphith ornamentation (Nikolov, 1999a).

According the the investigations undertaken in the recent years in Western Bulgaria and, in particular in the basin of Struma River, showed that as the end of the Early Neolithic is accepted the phase with polychromatic ceramic. (Chohadjiev, 2000).

In this period were established different cultures in regard mainly of their ceramic complexes. According to V. Nikolov (1999) one of them is the Variant "Galabnik – Sapareva Banya" which comprises the Neolithic culture alongside the upper and lower run of Struma River.

The first stage of the Early Neolithic is represented by the group Galabnik, characterized by vessels with red angoba and complex drawn in white ornamentation, along with Kremenik II and Balgarchevo II. The second stage of the Late Neolithic is registered in Kremenik II and Balgarichevo III (Nikolov, 1999a).

The transformation of the Neolithic ceramic of the valley of Lower and Upper Struma includes two basic periods and in it comes close to the variation Kapitan Dimitrievo. In a similar way the Early Neolithic period of the variant Galabnik – Sapareva Banya should have to be synchronized with the phases Karanovo I, II, II-III, III and III-IV in Karanovo and the late Neolithic period – with the phases of Karanovo IV.

SETTLEMENTS	DATE	AUTOR
KOVACHEVO	early neolithic	Lichardus, M. 2000
GALABNIK	early neolithic	Chochadjiev, S. 2006
TOPOLNITZA	late neolithic	Koukouli-Chrissantaki et all.
VAKSEVO	end of the late neolithic	Chochadjiev, S. 2001
BALGARCHEVO	end of the late neolithic	Pernicheva, 1995
DRENKOVO -PL.	late neolithic	Grebska-Kulova in press
SLATINO	early eneolithic	Chochadjiev, S. 2006
KAMENSKA CUKA	Late Bronze Age	Stefanovich, M. and A. Bankoff,1995
KOPRIVLEN	Late Bronze Age	Delev, P. and E. Bojkova, 2003

Fig. 19. Dating of the studied settlements in the region of the Struma River.

As result of series of archaeological studies concerning the Early Neolithic in the zone of the Central Balkans in the last years V. Nikolov defines three following groups:

Group Galabnik – the group has a continuous development in the frames of the first half of the Early Neolithic

Group Slatina – it covers the central and the eastern part of Sofia plane. Here refer the villages of Kremikovtsi and Slatina.

Group Gradeshnitsa – Karcha – it is characterized by the ceramic findings in the village of Gradeshnitsa

Group Kremenic – Anza – Bregovo – it covers the areas around the upper run of the Struma River

Group Kremikovtsi – it is located in the planes of Sofia and Zlatitsa-Pirdop

On the other side the development of the culture Karanovo III (its south-western variant) in the valley of Upper and Middle Mesta River goes through three stages – Eleshnitsa, Rakitovo and Dobrinishte (Nikolov, 1999a)

According to Nikolov today the Struma River is accepted as the main transmission for the arrival of the Neolithic to the Balkan Peninsula (Nikolov, 1999). The dating of the sites is presented in fig. 19 (after Popova, Marinova 2008).

RECONSTRUCTION OF THE NATURE OF AGRICULTURAL PROCESS

The studied settlements comprise the period Early Neo-

lithic – late Bronze Age and, to a major degree they are located alongside the main stream and the feeders of the Struma river (fig. 1). In the present study an attempt has been made to summarize the results of authors of several studies of sites namely – Kovačevo, Gălăbnik, Balgarčevo, Drenkovo-Ploshteko, Slatino, Vaxevo, Topolnitza, Kamenska Čuka, as well and to compare the data with that of some other scholars who have also studied sites in the region.

The composition of the findings of species cultural and wild growing plants is presented in fig. 20. The hulled wheat has prevalence in all sites that we studied. Also relatively well represented are the both barley species. It is noticeable that the barley or the einkorn is prevalent in the settlements located at a higher altitude (above sea level) and also in those of less favourable climatic conditions – as in Kremnik (Chakalova and Sarbinska, 1986), in Vaxevo and Rakitovo (Chakalova, Bojilova, 2002). The naked wheat appear sporadically which comes to show that in this period they have still not succeeded to settle. The seeds of the leguminous are of special interest.

In most of the settlements they were found in big quantities as storages. The availability of such a diversity is in connection with overcoming any risks that could have arisen from the yield – as some of them have higher resistance to drought and poor soils (*Lathyrus sativus* – grass pea and *Vicia ervilia* – bitter vetch), and both the peas and the lentils need higher degree of humidity. On that basis we could draw a conclusion that they have been of major importance as protein sources for man in the Neolithic. Most settlements are located in by-mountainous regions and near to water sources that provides good conditions for growing leguminous which need systematic provision of water. The finding of chick pea among the other leguminous is of special interest as, untill recently, it was known only in

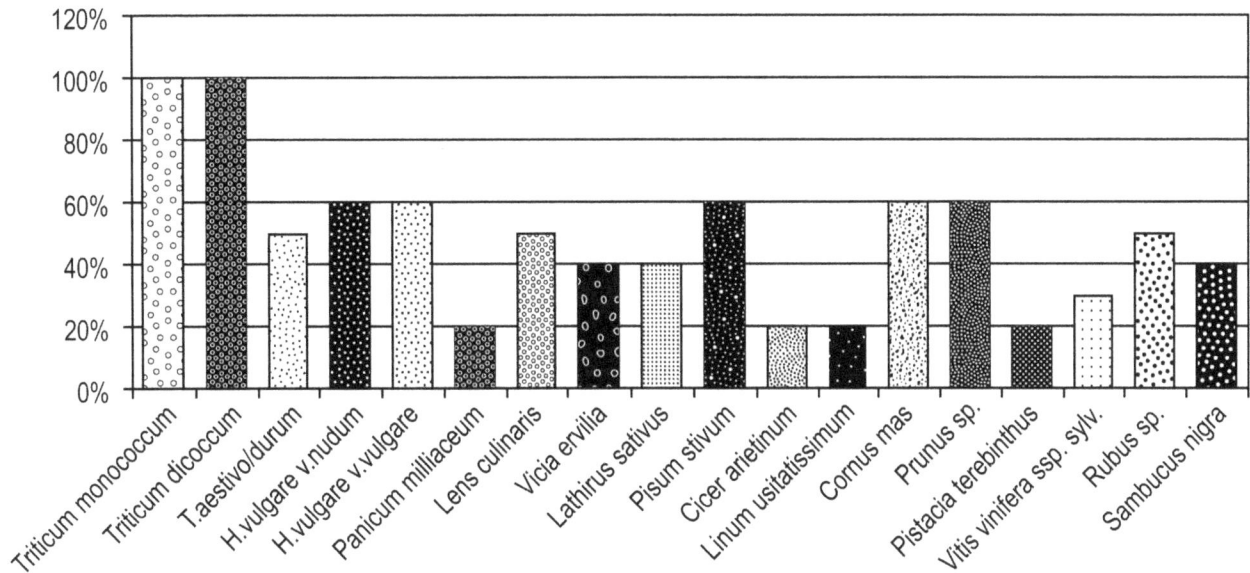

Fig. 20. Frequency of finding of cultural and wild growing plants in the studied Neolithic settlements in the region of the Struma River.

the southern parts of the Balkan Peninsula, and namely in Dimini and Otsaki (Kroll, 1981). This leguminous species comes to our lands from the complex agricultural plants of the Near East but before to succeed to get there equal position to the other cultivated species.

The wild growing plants had a specific part in human diet in the neolith. Often found in the studied settlements are a given number of fruit remains: of cornel cherries (*Cornus mas*), of *Sambucus nigra* - blue elderberry, *Vitis sylvestris*-wild grapes, *Rubus sp* - raspberry and partly of *Pistacia terebinthus,* which have been the subject of collection and the latter most probably used in the tanning of leather.

The situation of the seeds of the wild grassy and leguminous plants which are found in significant quantity in some early neolithic sites is different. Various wild growing plants could have been collected for consumption by man but their simple presence in the archaeological deposits is still not proof for their use in human nutrition. They could have been brought to the settlement also unintentionally (van Zeist, 1988). A good example is the available storage of sweet grass - *Chenopodium sp.* that appears in the chalcolithic site of Slatino. The presence of einkorn provides evidence that back in the neolith it had already been spread everywhere.

The recover of einkorn in the territory of Bulgaria are numerous (Hopf, 1973; Renfew, 1973; Popova, 1995). In the Balkan Peninsula it is present in the Neolithic epoch in the territory of Greece – Franhthi, Ahileon, Sitagroi II (van Zeist, 1980; Kroll, 1981; 1983); in the lands of former Yugoslavia – Starchevo; Obre I; Anza I-II (Renfrew, 1979); in Turkey – Cajönü (7500-6500 B.C.) (van Zeist 1972; 1988). The data of these findings prove that the Anatolian center even in the Mesolithic and especially in the early

phases of the Neolithic had been in close contacts with the Greek territories and in general with the Balkan lands.

The wide area of the emmer could be explained by its ability to develop under different ecological conditions. As there were limited amounts of arable lands, the type of the sowings had been of decisive importance for the preservation of the yield. The spikes of emmer do not fall apart, its steps are extremely strong and due to it they are used also for roof construction and for strengthening the walls. The plant is drought-resistant and it could be cultivated in different types of soil thanks to its well developed root system.

The barley in its two species appears in the Neolithic but then it occupies the second place with the exception of the settlements of Vaxevo and Rakitovo. It is obvious that the different barley species had entered in the territories of the Balkans through Asia Minor and Greece and that it was a process which took part in the Neolithic.

New data from Northern Greece (Valamoti, 2004) from the Neolithic layers of Dikili-Tash and Arkadikos show corresponding data to those of Middle and Upper Struma. In these settlements the prevalence of hulled wheat as well as a rich spectrum of leguminous was also established. Such predominance in growing leguminous plants should have been connected with their species with those resistance to drought: *Lathyrus sativus* – grass pea and *Vicia ervilia* – the bitter vetch.

The spectrum of the wild growing fruit bearing trees is represented by: apple, pear, etc., most probably preserved as dried fruit.

One gets the impression that archaeobotanically mostly

the neolith sites in the valley of Struma river were studied. This is mainly connected with the importance of the region in the clearance of the neolithic agriculture issues in our lands. Thus till now only one settlement from the eneolith epoch and one from the bronze epoch have been studied in the region. That suggests that future work in the region should also target the study of those later periods. And, in so doing, it will contribute to the reconstruction of a more entire picture of the ancient agriculture in the process of its development in the pre-history in the valley of the Struma river.

NOWADAYS VEGETATION AND RECONSTRUCTION OF THE FOREST WOOD VEGETATION

The presence of wood in the archaeological sites is usually connected with its usage in different human activities: construction, elaboration of production instruments, fire, etc. Almost always the wooden material was collected in the vicinity of the site and so the trees provide precise information about the flora of the studied region. Such studies are extremely important for our country, especially those in the valley of the Struma River where in the settlement of Galabnik was found exceptional diversity of wooden material.

MINERALIZED, NOT-CHARRED AND CHARRED WOOD FROM THE VALLEYS OF THE STRUMA AND MESTA RIVERS

The wood from this region could be classified in three different groups:

Mineralized – Topolnitsa settlement – there but was found not charred wood as well

Not charred – Galabnik settlement

Charred wood – in all the other studied settlements

Mineralized wood from Topolnitsa settlement

The settlement of Topolnitsa is dated from the Early Neolithic. The materials originate from the second horizon of dwelling No 2. The wood is strongly mineralized. The material represents a wooden awl, which is made from oak tree and a wooden needle made from wood of *Rosaceae* family species.

Defined are 3 charred fragments of pine trees, 2 of *Pinus sp.*

Not Charred Wood from Galabnik Settlement

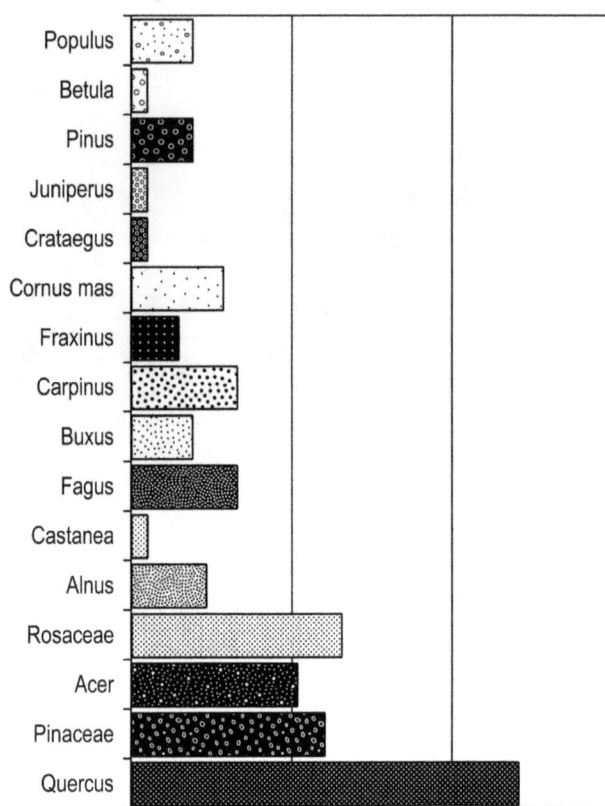

Fig. 21. Distribution of mineralized wood of the Neolithic of Galabnik (according to Grozeva, not published data (Popova, Marinova, 2008).

A tell settlement found by the correction of the river bed of the Blato River, left feeder of the river of Struma River. The area studied there is of 1300 sq.m. There were found 10 early Neolithic horizons: 7 of the culture Proto-Starchevo and 3 of the culture Starchevo (I horizon is the earliest one). Initially the region was dry but quite soon it changed, most probably due to erosion caused by the fast cut of the local forests, and the level of the subterranean waters rose significantly higher. In the first two horizons no wood was found but in the III and IV horizons the stakes of the dwelling bases, separate wooden boards, and big fragments of wood, some of them together with their bark and gemmae were preserved. The small fragments are silicized. Recently 1,70 m of the cultural layer is under the water level and for that reason only an area of 50 sq. m – was stratigraphically studied. 21 wood species were found in a total of 126 fragments. The predominante are the oak fragments – 26, followed by those of *Rosaceae* species – 14; ash tree– 15; pine - 13. It was established that the ashl tree three was used for the longitudinal planks, and the wooden material of pine species, Norway maple, and *Rosaceae* – for elaboration of subjects used in the everyday life.

The non charred wood in the settlement of Galabnik (Fig. 21) was studied by Grozeva and in consequence docu-

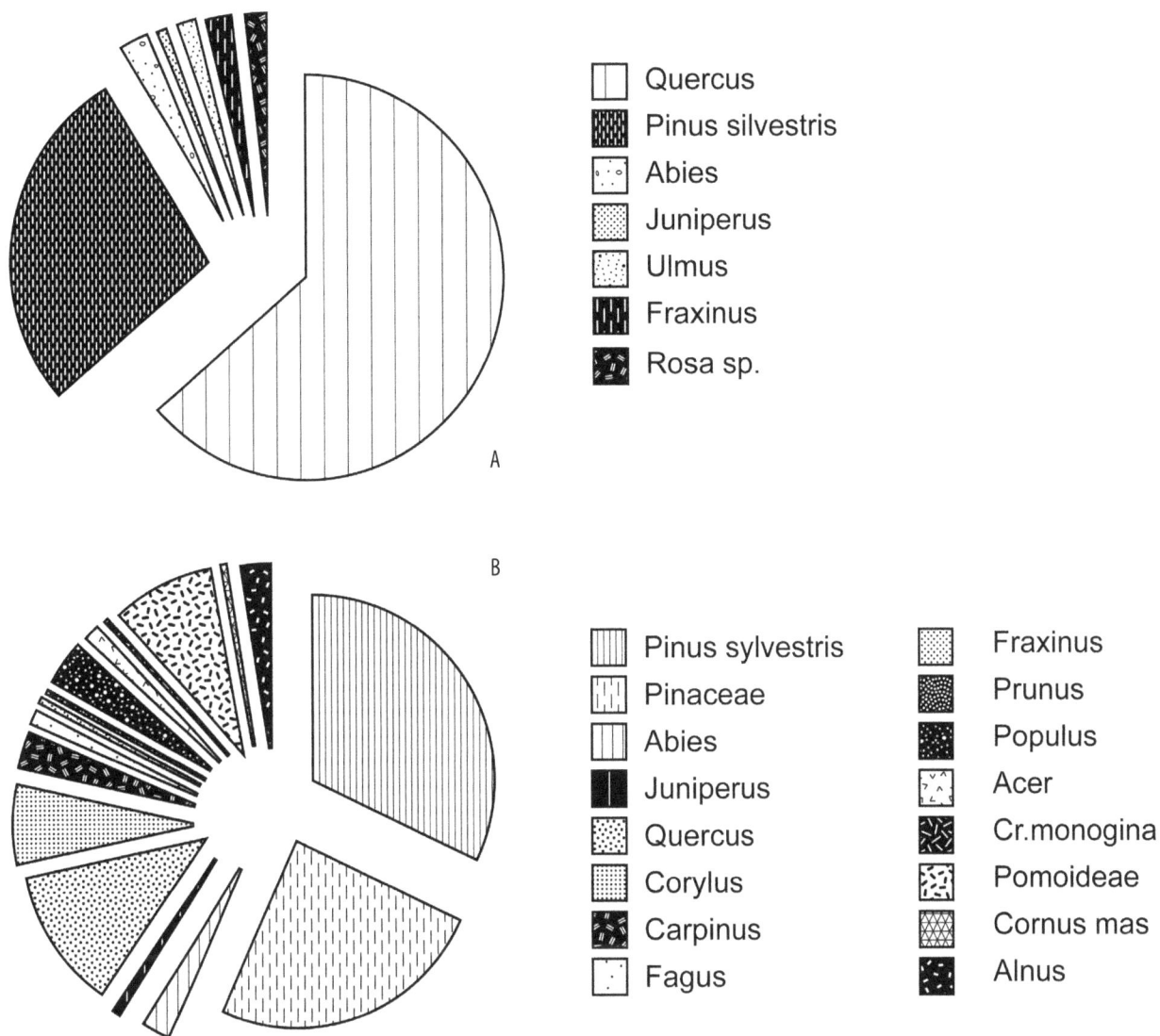

Fig. 22. Charred wood from Late BronzeAge settlement Kamenska cuka (A) and Koprivlen (B).

mented by E. Marinova et all., 2002c.

In the Neolithic settlement Kovachevo together with the predominant oak (*Quercus sp.*) is also widely spread the cornel tree species (*Cornus mas*). Possible explanation of its strong presence could be its usage as construction material. The cornel tree has robust and durable wood which is extremely suitable for the entwine which has been of use for walls of the houses.

The rivers located alongside the rivers were of *Alnus* - alder, *Ulmus* – English elm, *Fraxinus* – ash tree, and the sub-Mediterranean flora was represented by *Pinus nigra* – Austrian pine.

Oak wood was found also in the site of Kamenska Čhuka. (Fig. 22.A) (Popova, 1999a).

In smaller quantities were found small parts of: *Abies sp.*;

Juniperus sp.; Pinus sylvestris; Pinus nigra; Ulmus sp.; Fraxinus sp.; Rosa sp.

It is possible that they were collected in more remote locations from the settlement places as well as from the higher places in the surrounding mountains (Popova, 1988).

Charred oak wood is documented in the sites of Vaxevo and in Koprivlen – Fig. 22.B.

Different oak species were wide spread and their role in the archaeological context could be explained by some of their qualities. The oak wood is very strong, the oak could be found in easily accessible places as the settlements were located by such height above the sea level that they were in the habitat of the mixed oak forests. By burning they reach high temperatures fast which is highly preferable for the households.

Charred wood of Rakitovo pre-historic settlement was studied by Chakalova and Bojilova (1991).

The studied fragments of charred wood in some of the settlement alongside the Struma River provide opportunity for some conclusions to be drawn as well as allow comparisons to be undertaken about the type of use wood and the altitude of the settlements.

It is established that in all studied settlements the wood collection took place in the altitude of 150 to 1500 m. Still it should be taken into consideration also the wide diapason of distribution of some species as *Abies sp, Fraxinus sp. Pinus sylvestris, Acer campestre* etc. It is obvious that the wood collection was preferred as activity to take place in the lower places (by a lower altitude).

In the period around 8000 B.C. according to the climate reconstruction done on the basis of the precipitation of saprophyte in the Mediterranean regions (Davis et al. 2003) most likely there was an increase in the humidity in Soth-Eastern Europe. The palinological data of Pirin Mountain show warmer and more humid conditions in the period around 7200-6500 B.C. At that time the broad-leaved deciduous forests of the moderate climate reached habitat in higher altitude than that of today. According to the palaeoecological studies in Pirin Mountain there was a change in the seasoning of the climate in the period around 6000 B.C. when the summers became considerably cooler and the winter considerably warmer (Stefanova and Amman, 2003).

According to the data published by Bojilova and Chakalova (1981) for the settlement Rakitovo in that period the pine species spread to an altitude of 1000 m. The found tree species are typical for well-drained soils. The found tree species are typical for well drained soils. Predominant is the oak wood and the species most often found are: *Quercus dalechamii; Querus polycarpa*. The wood which was found could provide information about the palaeo environment but with some reservation. In the settlement of Rakitovo is found hazel species. It grows fast and develops as a first pionner and after the clearing of forests it could be found in close vicinity of the settlements. In addition it is used also for fences.

The English Elm tree is one of the tree species that grows in the lower part of the forests as it requires higher humidity. The English Elm tree develops in conditions of moderate continental and more humid climate. This tree species prefers deeper and humid soils. It enters into the composition of the mezophil mixed oak forests alongside the rivers. The authors compare their data with pollen analysis that provides the following picture: the data show that the cultural layers of the settlement refer to the so-called Atlantic period (8000-5000 B.P.). It is characterized with a general increase in the temperature and in the rain falls

which, by that altitude caused increase in the forest areas, migration of the tree species to the mountainous slopes resulting in a rise in the upper limit of the forests.

The *Pyrola* fruit found shows that the forests around the settlement of Rakitovo were shadowy, with chady, coniferous trees, and with significant humidity in the soils and in the air (Chakalova and Bojinova, 1981; 78).

The contemporary vegetation in the region consists of Scots pine, and upper follow oak forests and mixed forests of beech and silver fir, depending on their location to the sun.

The pre-historic settlements in South-Eastern Bulgaria are rich in archaeobotanical material. On the basis of the findings – as species, as frequency of finding of remains and dependent on their quantity as well as on their comparing with the spread of those tree species nowadays some conclusions about the former flora and vegetation in these lands could be draw as welle as about the climatic changes in the region.

The wide spread of the einkorn - *Triticum monococcum* - is based on its ability to get acclimatized to different ecological conditions. Thanks to its well developed root system the *Triticum monococcum* develops well even in unfertile mountainous soils as well as in extremely humid soils.

The emmer - *Triticum dicoccum* is represented in the studied settlement in big quantities.

Barley occupies the second place among the crops in the studied samples. Additionall to the wheat cultures a wide spectrum of leguminous were used also. So a conclusion could be made that the representation of these species in the studied settlements in the territory of Bulgaria is comparable to the results of other settlements in the Balkan peninsula.

The wooden material found in the archaeological layers is evidence of the predominance of oak forests in the region as well as for the usage of the diverse ecological niches in near vicinity. The collected tree species in the river valleys were of: *Ulmus sp, Alnus glutinosa/incana, Fraxinus sp.* and fruit trees. The data about the use of Austrian pine (*Pinus nigra*) give evidence that this species were much wider spread in the past than nowadays by low altitude and in limy lands.

2. THRACEN-STRANDJA SUB-REGION

Physicogeographical characteristic of the region

The Thracian-Strandja region is situated between the mountain range of Sredna Gora and the Rilo-Rhodopes

Mountain massive and, to the south-east it ends at Bulgaria's border with Turkey. To the east it reaches of the Black Sea. The region comprises the Upper Thracen plain, the Middle-Tundja River valley, the Sakar and Strandja mountains, the Sveti Ilijski, Manastirski and Derventski hills and the Haskovo hill, slightly transitional to the Eastern Rhodope Mountain. The Upper Thracen lowlands cover vast area between the Sredna Gora Mountain ranges and the Rhodope Mountain. The nowadays landscape of the Upper Thracen plain and of the Middle-Tundja valley was formed in the Quaternery after the drying of the lake basins. The Upper Thracen plain divide into two parts – Plovdivsko-Pazardjishka plain and Starozagorsko plain. The boundary between the two plains is between the Chirpan hills and the Rhodopes hill Dragoyna.

Important climate formatting significance in the Thracen-Strangian sub-region is proximity to both the low altitude and the Black Sea. The climate in the Upper Thracen lowlands is transitional-continental. The winters are comparatively mild (average January temperature 0° - 1°C), and the summers are hot (average July temperature 23° - 24°C). The average rain falls are lowest in the western part of the Upper Thracen lowlands (500 - 550 mm). Predominant are the western winds. Sakar and Strandja mountains are more appreciable under the influence of the Mediterranean. The climate there is Continental Mediterranean.

The basic draining roads in the Thracen-Stranjian sub-region is the Maritsa River. Its waters are collected mainly from its Rhodopes feeders the rivers Chepinska, Vatcha and Chepelarska which are significantly more deep watered than its feeders coming from the Balkan Mountain – the rivers Topolnitsa, Stryama and Sazlijka. The river of Tundja enters its mid steam to the north of the town of Yambol where it merges with its most significant feeder – the river of Motchuritsa.

Most widely spread soils here are the resinous lands the alluvial-meadow and the cinnamon forest soils. Because of their high natural fertility the soils here are favourable for intensive agriculture. Many crops are grown here, such as: wheat, maize, rice; technical cultures – sunflower, cotton, tobacco as well as also vines and fruits.

The natural vegetation in the Upper Thracen lowlands and in the Middle Tundja by-river lands is almost entirely replaced by cultural plants. Here and there among the arable lands there are still separate zones of bushes and trees communities, consisting of deciduous species (*Quercus pubescens, Carpinus orientalis, Paliurus spina christi*). In the low mountains of Sakar and Strandja predominant are Mediterranean plant species such as *Jasminum officinale, Asparagus tenuifolius, Erica herbacea* etc. The vegetation inside Strandja Mountain is of relict character. It consists of forest communities, formed by *Fagus orientalis* accompanied by the ever green *Rhododendron ponticum* – the

special Strandja species, *Prunus laurocerasus* etc. Such type of mountain vegetation in Europe is found only in Strandja Mountain.

The physicogeographical conditions of this region define the preconditions for the development of economy as early as in the Neolithic – one of the biggest rivers of the region, the Maritsa River runs through the region, the soils here are fertile and the climate is mild.

Chronology

A big part of the Neolithic and the Chalcolithic settlements here are defined as belonging to the cultures Karanovo I – IV and other refer to the Bronze Age.

The periodization of the Neolithic in Thrace was suggested 40 years ago by G. I. Georgiev (1974). It is developed mostly on the basis of his own observations by the excavations of the tell settlement Karanovo, Nova Zagora municipality and later completed after the excavations of the tell settlement Azmak by the town of Stara Zagora and of the tell settlement by the town of Kazanlak. According to this periodization and chronology the Neolithic culture in Thrace have 4 basic stages which were established for the first time in the tell settlement of Karanovo and named after it: Karanovo I, II, III and IV. Later the system was supplemented with the period Karanovo II – III which Georgiev defined in the tell settlement by the town of Kazanlak. According to Georgiev (1974) those 5 periods followed one after the other in the entire territory of Thrace. The first two cultures – Karanovo I and Karanovo II are referred to the Early Neolithic, Karanovo III – to the middle Neolithic and Karanovo IV – to the late Neolithic Nikolov (1998).

Doubts about the validity of that periodization for all Thracen lands appeared after the excavations done in the tell settlement by Azmak and in Kazanlak where no traces of the period Karanovo II were found. Similar problem appeared also in other Neolithic settlements in the western part of Thrace – Muldava, Assenovgrad area, the village of Banya, Karlovo area, Rakitovo, Capitan Dimitrievo, Peshtera area, etc. These are the main preconditions for the understanding that Karanovo II, which belongs to the second half of the early Neolithic developed only in the North-Eastern part of Thrace and thus Nikolov accepted the name 'group Karanovo II'. Nikolov (1998, pp. 16-17). According to the pottery complexes a variation 'Karanovo' is determined, which consists of several layers:

Layer Karanovo I – it comprises 4 building horizons. The basic characteristic of the pottery are the vessels with red angoba and sometimes ornamented in white color. This layer refers to the Early Neolithic in Thrace.

Layer Karanovo II – it includes 4 building horizons and it refers to the Early Neolithic.

Layer Karanovo II – III – here are included 2 building horizons. The pottery has almost the same typical peculiarities of the previous periods but two new kinds of vessels appeared. This layer refers to the Middle Neolithic in the North-Eastern parts of Thrace.

Layer Karanovo III – it comprises 3 building horizons. The pottery already includes the several basic vessels. This layer refers to the first half of the Late Neolithic in Thrace.

Layer Karanovo IV – with 3 building horizons. Refers to the second half of the Late Neolithic in Thrace.

The studies in the region of Sazliyka River provide basis to the Panayotov, 1991 to presume the existence of a central settlement with situated in its vicinity other detached open air settlements. As for example is the case with Ovcharitsa II and Golyama Detelina. A similar opinion expresses also Leshtakov, 1992: "… During the Early Bronze Age 2 appeared the so-called "satelite-settlements". They are documented in Yunatsite and in Mikhalich…"

In addition to these known variations there are also variations of Kazanlak and Kapitan Dimitrievo, which are typical for Thrace (Nikolov, 1999).

MARITSA – IZTOK

Maritsa Iztok region not less important than the others the region of the plain of Maritsa River and, in particular there are concentrated a series of archaeological sites comprising such of the periods from the Neolithic to the Middle Ages. Some of them are archaeobotanically studied. In the present study I include just those which refer to the Prehistory. The Maritsa Iztok Complex is the largest energy complex in South Eastern Europe; however, the mining activities cause grave ecological problems: destroy of the fertile upper layer of the soils, air pollution, the waters and the soils with toxic substances. As result of the landscape changes and consequently the changes in the hydrological conditions due to which here are the biggest landslides. Demolished are vast areas of arable lands - especially around Assenovgrad, Galabovo, Radnevo and Dimitrovgrad – they are contaminated by heavy metals which turns them into unfit for agricultural use. As basic contaminator there appears the huge electricity producing station.

Part of the studied settlements in this vast region is connected with rescue excavations: Mednikarovo, Galabovo, Madrets, Golytama Detelina, Iskritsa, Polski Gradets. Archaeobotanically studied are the following settlements: Karanovo, Dabene, Chatalka, Azmak, Nova Zagora, Yunatsite, Yazdach, Nebet Tepe, Galabovo.

The studied settlements cover the period of the Neolithic till the Bronze Age.

In their bigger part these settlements represent tell settlements which, for Thrace, is a typical feature (Leshtakov, 2002).

RECONSTRUCTION OF THE NATURE OF AGRICULTURAL PROCESS

The archaeobotanical studies in this region comprise mostly field studies and rescue excavations. By the study of the archaeological sites was done interdisciplinary research which aimed at the reconstruction of the palaeoeconomy and of the influence of man over his environment.

In that connection much better success is achieved than that of the earlier palaeobotanical studies of the tell settlements near Ezero, Nova Zagora and Chavdar (Dennel, 1972; Hopf, 1973). Later some other settlements in the same region were studied. The contemporary landscape of the region – and in particular alongside Maritsa River – is characterized by low hills, slight slopes and plain lands. Running through the region is Sazliyka River which springs from Sredna Gora mountains and its feeders. In the river valleys there is an interesting natural phenomenon – mud volcanoes. The nowadays landscape though gets changed due to the erosion.

The climate is transitionally-continental and here the influence of the Mediterranean climate in the district is too weak. Typical for this region is a hot summer and a comparatively mild winter. The average minimal temperature is of 14 degree C and the average maximum temperature is 35 degrees C. The soils of the river flooded terraces are of alluvial and meadowy character; there are also black earth soils. The contemporary vegetation consists of strongly degraded formations of hornbeam - *Carpinus orientalis*, oak – *Quercus sp.;* bushes - *Paliurus spina – cristi*; grassy vegetation resistant to drought. The influence of the industrial activity of man in the region is strongly felt.

The studied material represents charred grains of cultural wheat and leguminous as well as fruits and seeds of wild growing species. They are collected through flotation.

The studied material provides basis for drawing conclusions about the composition of cultural plants in the studied region, sufficiently varied. Several species of hulled wheat were grown but the dominant culture was the einkorn. Its biological features could explain the wide spread of this species in the Balkan Peninsula and its prevalence over the other wheat cultures. The species is resistant to unfavorable factors of the environment, to unfertile rocky soils as well as to the cold weather and to drought. Data from other archaeological sites also provide evidence that this corn is less demanding. Its yield is not high but it is quite consistent.

As concerning the barley, the analysis shows that by similar ecological conditions in different periods and cultures from the Neolithic to the Bronze Age the barley was one of the preferable corns. And, taking into consideration that the plant gets self-sown and very often spreads as weed among the sowings of other wheat cultures, it should not surprise us. Concerning the sporadic findings of spelt – *Tritium spelta*, obviously the climatic conditions for it were to some extent less favorable (more dry climate and/or infertile soils) as it was less often sown.

The naked barley was preferred than the hulled species because it is easier for growing as by it there is no need to separate the grains of their cover. Additional to it this wheat species is more resistant to unfavorable conditions.

Among all other leguminous the bitter vetch is, no doubt, of special interest. It is found in all studied sites. It is not clear yet why this leguminous species was preferred than the lentils and the peas (the elimination of the toxic substances from the bitter vetch through stay in water makes the preparation of food of it a longer process). Principally the complex of vegetable species shows the existence of corns which are able to get adapted to different kinds of soils and to sustain by different climatic conditions. All cultural plants show similar structure also in the studied sites of: Dyadovo, Nova Zagora, Plovdiv, Yunatsite (by Pazardjik), Galabovo (by Nova Zagora) Popova (1991b, 1991c; 1995b), (Popova and Bojilova 1983). The archaeobotanical data concerning the Neolithic agriculture achieved by Marinova for the settlements of Kapitan Dimitrievo and Karanovo show similar results. Most studied as quantity materials by her are the food plants– the wheat. As basic wheat plants appear to be the einkorn and the emmer. Often found also, though in smaller quantities, is the barley.

Object of collection in the Neolithic were the cornel-tree, cherries, elder, raspberries, blueberries, plums, strawberries, common dogwood. The weed flora provides evidence that there were also species sown in the autumn. Some of them were with a maximum height of growing 30-40 cm.

The analysis of the flint assemblages from Southern Bulgaria show the availability of big quantity elements of sickles as well as the fact of its continuous usage – documented by the degree of their usage. It gives the basis for Gurova, 1999, to assume the presence of a highly developed agricultural practice and tradition still in the early phase of the Neolithic in Thrace. The presented observations provide evidence for the presence of settled agriculture with developed producing economy (Gurova, 1999).

Nowadays vegetation and reconstruction of the forest wood vegetation

The presence of charred wood gave me the possibility to study wood in almost all of the studied archaeological sites. Of special interest though is the stratigraphic study of a profile from the tell settlement Madrets as well as from the building horizons of the tell settlement in Galabovo. As summary for those two settlements as also of the data that we have for other archaeological sites neighboring the studied region and of synchronized dating, we could draw the following conclusion: The analysis of the charred wood provides evidence that the oak was the dominant species. Certain formations as live fences also existed with the participation of some fruit bearing trees (plums, cherries, cornel-tree) and ash tree and English elm tree were also to be found. The alder and the hazel were growing in the cleared spaces and alongside the rivers. Most of the found tree species are heliophilic and moisture loving species The English elm habitat was in the lower parts of the forest where there was more humidity. Often met was also the hornbeam that prefers the moderate-continental climate. It grows in crumbly dilluvial soils.

The palaeological studies concerning Thrace and Western Bulgaria provide evidence for the extensive usage of oak forests during the Bronze Age. It was connected with the clearance of forests for pasture lands and for growing cultural plants. The forests of English elm and hornbeam were also destroyed as the settlements climbed up the mountainous slopes. Many found ruderal plants witnessed of intensive cattle breeding during the Bronze Age.

In the low pre-mountainous lands the conditions for development of the settlements were favorable. Significant change took place with the destruction of the pine forests and increase in the beech communities. In the high mountainous lands in the Bronze Age there was the practice of season grazing, the forests of Pinus mugo were often put to fire. At the end of the Bronze Age the significant presence of wheat, barley and rye in the high mountains in our lands are marked in the pollen diagrams. The increase of *Artemisia, Plantago lanceolata, Rumex, Cirsium type* are connected with the clearance of forest territories for pasture lands Bojilova and Popova (1998).

According to data provided by Stefanova and Philipovich (1997, pp.103-116) in this period in the mountain range of Sredna Gora were dominating the forests of hornbeam, spruce, hazel, birch, followed of a beech with admixture of pine trees. Mixed broad-leaved forests covered significant areas in the valley of Tundja River.

3. Northern Bulgaria

Physicogeographic characteristic of the region

In this part of the country are studied settlements some of which are located in the plain of Danube River, and more precisely in its Eastern part. The settlements are dated

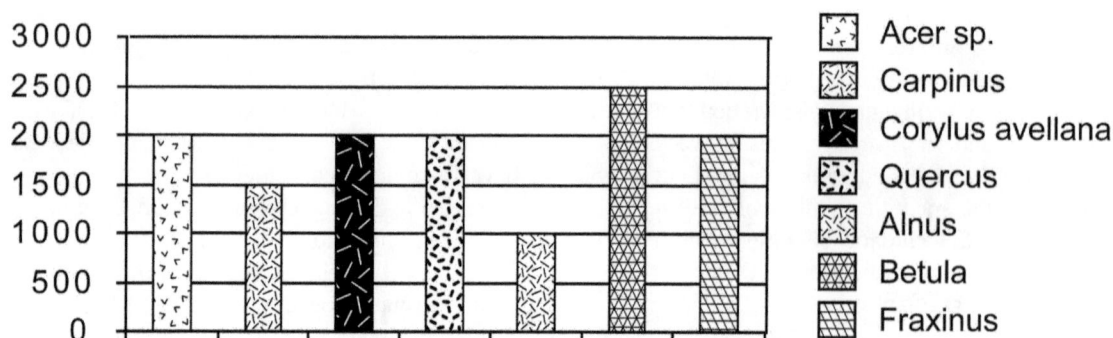

Fig. 23. Dissemination of the found forest species according to the height about the sea level in the region of the region the Maritza River region

from the Early Neolithic to the Bronze Age. They are as follows:

Neolithic – Drinovo, Malak Preslavets, Ohoden

Middle – late Neolithic – Podgoritsa

Early – middle – late Neolithic – Orlovets

End of the Chalcolithic – the second half of 5 000 B.C. – Omurtag, Suvorovo

Chalcolithic – Hotnitsa

Chalcolithic – Bronze Age – Durankulak

The Danube plane is situated between the Danube River to the North and the pre-mountainous hills of the Balkans to the South. To the West it spreads to the Timok River and to the East reaches the Black Sea. The Danube plane is formed on the Misia plate which is characterized with flat layered relief. The lowlands alongside Danube are filled with alluvial clay-sandy and gravel –sandy sediments aged from the Quaternary.

The average height above sea level of the Danube plane is 178 m. The Danube plane has plane and undulated land and plateau relief. It is divided into three parts: western (from the River of Timok to the River of Vit), middle (from the River of Vit to the River of Yantra) and eastern (from the River of Yantra to the Black Sea).

The moderate-continental climate of the Danube River is formed by its considerable openness to north-east and its considerably unvaried relief. Its average annual temperature varies from 10°C to 12,2°C. The average January temperature for the plane is about -1°C. The highest summer temperatures are in the month of July - 23°-24°C. The rainfalls are about 600-650 mm. Predominant are the western, north-western and northern winds. The soils in the plane are in their bigger part formed on loess basis by availability mostly of steppe and forest-steppe vegetation. Developed there are mostly the black earth "chernozem"

soils (carbonate, typical, etc.) and less grey forest soils. The soil types and the specific climatic conditions determine the transition from broad-leaved forest vegetation to the west to more dryly vegetation with steppe character to the east. The natural vegetation occupies today quite restricted areas (in the regions unsuitable for agriculture). On the basis of this vegetation it could be presumed that in the past the Danube plane was almost entirely occupied by vast forests and steppes. Especially covered by forests was the eastern part of the plane – the Ludogorie. Today the natural vegetation is preserved in the Danube islands and in the unused for agriculture parts of the lowlands directly by the river bed where then level of the undersoil waters is high. In the content of these forests are included mainly moisture loving - willow and poplar. From the tree species most spread are some oak species, hornbeam, English elm, hazel, and lime. From the steppe species are dominating the steppe wheat species and some *Rosacae*, etc. In the forests dominate species as: *Quercus pubescens, Q. virgiliana, Q. dalechampii, Q. cerris, Q. frainetto, Fraxinus ornus, `Acer campestre, A. tataricum , Pirus communis* L. *subsp. piraster* (L.), *Ulmus minor*. From the bushes often to meet are the following species: *Cornus mas, Colutea, arborescens, Syringa vulgaris, Viburnum lantana, Crataegus monogyna, P. spina christi, Juniperus oxycedrus*. In the grassy communities are: *Heleborus odorus, Brachypodium pinnatum, Brah. sylvaticum, Dactylis glomerata, Veronica chamaedrys, Ramonda sertica*. Nowadays the forests in the Danube plane are degraded to a big extent.

One part of the prehistoric settlements in is located in the western, other part in central and other in Eastern North Bulgaria. For that reason here a brief characteristic of the basic cultures is to be given.

According to Nikolov, B. (1992) in the Early Neolithic in the territories of the nowadays Bulgarian lands are defined 4 early neolithic ethno cultures which are named as follows: culture Karanovo I – for Thrace, culture Kremenik – Azanbegovo, which covers the valleys of the rivers Struma and Vardar, culture Kremikovzi for the Sofia, Radomir and part of the Kyustendil fields; culture Gradeshnitsa – from the river of Yantra to the river of Timok in Oltenia district of the south-western Romania (Nikolov, B. 1992). The found archaeological evidences provide opportunity the Neolithic period to be divided into three phases: early, middle and late Neolithic. The early Neolithic is characterize with the painting of vessels of fine pottery with different colour paints.

The middle Neolithic in this region, according to Nikolov (1992) is connected with a substantial change which he characterized with the absence of the fine polychrome ornamentation. The settlements are characterized also with their location in open air places near to water sources (springs). Proof of the presence of late Neolithic in the region from Yantra to Timok was evidenced only in the village of Brenitsa, Vratsa district.

Other parts of these settlements in north-eastern Bulgaria refer to the so-called culture Gumelnitsa – named after the tell settlement with the same name. The tell settlement Gumelnitsa is located near to the left bank of the river of Danube. It is dated from the beginning of the Chalcolithic – 5000 B.C. The culture has quite wide expansion along the Black Sea cost, to east and west in Central Bulgaria, from the Danube delta in the north to Greek Thrace to the south. Its settlements basically are tell settlements – Karanovo, Harshova, Bordushani and their stratigraphy provides enough information about the chronological evolution of these cultures and of their connection with the neighboring cultures – Vinča, Kukuteni, Salkutsa. Studies concerning the Bronze Age in Northern-East Bulgaria are connected with the study of materials – from Alexandrov, 1995; Alexandrov, 2003 where the typical culture is the culture Balley-Usoe.

From the culture Baley-Orsoya are known considerable number of settlements in north-western Bulgaria. The settlements of the culture Baley-Orsoya are situated in the valleys, almost always near to the rivers (Panajotov and Valcheva, 1989).

RECONSTRUCTION OF THE NATURE OF AGRICULTURAL PROCESS

The results of the archaeobotanical analyses of these settlements show the following picture: Concerning the quantity of the studied material there are certain differences in comparison with the other studied regions. First, here the settlements that we have studied are less in number. In most cases with the exception of the settlement Hotnitsa the extracted plant remains are in not significant quantity. That is due to reasons on which we do not have control. Still a certain picture of the grown plants could be established. In general hulled wheat is evidenced – einkorn and emmer. In some of the settlements is observed only a single type of barley – the naked species and in others only the hulled species or both (Panayotov, et all., 1985). Noticed is the

availability of common/durum wheat but it is presented by single grains. Of interest is the finding of several grains of spelt. This species principally is not sufficiently spread in the territory of Bulgaria in this period. The spelt is a wheat species which grows better by cooler climate so it migrated to the north where it has successfully acclimatized. In the most Neolithic and Chalcolithic settlements in Central and Northern Europe it is found much more often than in our lands. Interesting difference is observed also concerning the leguminous plants. In all settlements here are found representatives of the leguminous – lentils, peas, bitter vetch, grass pea and chick pea. In some of them they are even without admixtures and in big quantities which most possibly present storages. Such are the found storages in Omurtag as well as in Hotnitsa (Popova, 1985, Popova, 2008). In some of the cases they even dominate as percentage compared to the wheat cultures. That could be explained on one side with the ecologic peculiarities of the species and, on the other side with the climatic conditions in the region. It is known that the peas and the lentils are plants which prefer cooler climate and higher humidity which is a pre-condition for a better yield. In the case here the moderate-continental climate as well as the higher degree of humidity – the access to rivers or systematic irrigation – were favorable conditions for their development as well as for that of the spelt. The found materials provide basis the following conclusions to be done:

Prevalent growing of leguminous which to some extent distinguishes these sites.

Presence of almost all species of the leguminous known and characteristic for the epoch.

Appearance of the wheat species spelt.

NOWADAYS VEGETATION AND RECONSTRUCTION OF THE FOREST WOOD VEGETATION

The study of the charred wood in these settlements is insignificant. It is so because in most of the settlements the samples were collected accidentally by the archaeologists and no flotation was done. Concrete data are available only from the settlement of Hotnitsa which show the following: in the most cases the most used species is the oak. The presence of ash tree in the region could be connected with the existence of more humid conditions as it develops well in cooler places in soils with constant humidity. Its wood as the found wood of maple is very suitable for the elaboration of small carpenter tools/ instruments. By presumption could be suggested that also in the other settlements the oak was the main source for construction.

The contemporary vegetation in the region is represented

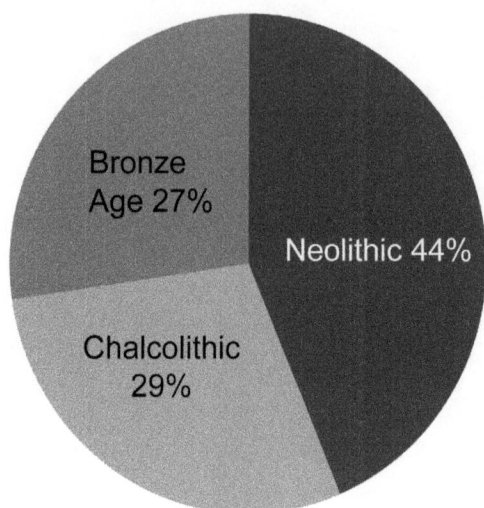

Fig. 24. Quantitative distribution of the Triticum
monococcum –einkorn in the studies settlements from
different periods.

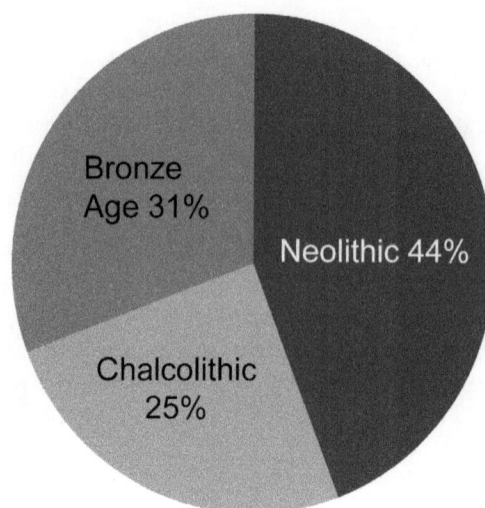

Fig. 25. Quantitative distribution of the Triticum
diccocum –emmer in the studies settlements from different
periods.

by mixed oak forests of *Quercus cerris* L. *Q. pubescens* Willd, *Q. virgiliana* (Ten.)Ten), grassy vegetation with domination of *Poa bulbosa* L., *Lolium perenne* L., *Cynodon dactylon* (L.) (Pers), *Dichantium ishaemum* L. and less often of *Crysopogon gryllus* L.(Trin). Except them are also to be found *Tilia tomentosa* Moench, *Carpinus betulus, Acer campestre.*

As evidence for the availability of oak forests as well as for the presence of the rest tree species could appear the pollen studies of the lake of Srebarna natural reserve. In Danube Dobrudja then balanced mixed xero and mezophyle oak forests dominated by different oak species. Important elements in the composition of the forest were *Carpinus betulus, Tilia, Ulmus, Acer, Corylus, Fagus.* Communities of *Alnus, Salix and Populus* were formed alongside the rivers, around lakes and in some other humid places. The presence of pollen of cultural wheat plants and the typical indicators for pastures – *Polygonum aviculare, Plantago lanceolata, Centaurea cyanus, Rumex, taraxacum type* prove the availability of arable lands. According to the results of the pollen analysis in the early Chalcolithic – 4800 ± 110 B.C. or 6800 - 5000 B.P. the lands around the lake was covered by mixed oak forests and there was no significant difference in it from the Neolithic. Probably the arable lands space had increased which was documented by the increase in the pollen of *Centaurea type* (Lazarova, 1995). The natural and geographic conditions in the Balkan Peninsula in the period 7000-5000 B.C. are characterized by significant warming in the climate accompanied by increase in the humidity (Bottema, 1974). There had been increase in the average annual temperature which had been higher that the contemporary one with 2-3 C. The favorable conditions brought increase in the human populations and further cleaning of places with fertile soils. The culmination in the Climatic Optimum is characteristic for the

end of the 8000 B.C. to the end of the 5000 B.C (calibrated C14). The archaeological data prove development of the cattle breeding in the district of Durankulak (Todorova H., 1989). Evidence of intense agriculture and economy is provided also by the archaeobotanical materials of the author collected from the district of Malak Preslavets, Durankulak, etc.

These data correlate well also with the data of the pollen analysis which establishes the presence of series of ruderal plants as: *Plantago lanceolata, P. media, P. major, Polygonum aviculare ,Centaurea cyanus.* The increase in the populations of the hazel and the hornbean could be explained with the mass cut of the oak forests by the Chalcolithic man and the appearance of secondary vegetation communities (Lazarova, 1995). The clearing of lands happened through fire and cut.

In the Bronze Age – non calibrated 5000 – 3500 B.P. (3000 – 1500 B.C.) and in the Early Iron Age which connects with the sub-Atlantic period – 3000-2500 B.P. in the pollen diagram from the national reserve of Srebarna lake the influence of man over the natural environment is established extremely fast. The tree species diminished drastically, the anthropogenic impact increased and there is a significant increase in the cultural wheat pollen. The oak trees are represented as local elements. The cleared terrains were occupied by communities of *Carpinus orientalis, Fraxinus ornus* which were found in the sun shined slopes. With the processing of the soil together with the sown plants appeared also the typical weed indicators which were established in the pollen diagrams (Lazarova, 1995. Archaeological, palinological and archaeobotanical studies in the region of the lake Varna (run in 1972-1979 by the archaeologist I.Ivanov and G. Toncheva – Toncheva, 1973) provide opportunity for answering the question about the type of agriculture, the utilities used and about the influence of the cattle breeding (Bojilova and Ivanov, 1985; Bojilova, 1986). In the region of Varna Lake are found a significant number of Bronze Age settlements – as for example Strashimirovo (Ivanov, 1973). The layer in the 'Arsennal' of the profile (400 – 200 sm) is dated 2730 +_70 B. P. – Early Bronze Age. In the settlement Eze-rovo are found charred grains of einkorn, emmer, hulled and naked barley as well as of bristle – grass (*Setaria sp.*) from the Early Bronze Age. The charred wood is in big quantities. It belongs to the following species: *Quersus sp., Acer sp., Ulmus sp., Fraxinus sp., Evonimus sp., Rosaceae (subf. Pomoidea)* and *Vitis sp.*

* * *

The analysis of the archaeobotanical material shows that in big part of the pre-historic settlements the basic plant species grown were the hulled wheat species – einkorn and emmer.

In figs. 24-25 is shown their quantitative distribution during the different periods. From there it could be seen that the presence of einkorn and emmer in the Neolithic is equal – 44%, in the Chalcolithic the emmer diminishes its presence – 25% for the emmer compared to 29% for the einkorn and in the Bronze Age it increases – 31%.

Chapter 6

AGRICULTURE.

Concepts for the Origin of Agriculture
on the Basis of Archaeobotanical Data

The archaeobotany is the science whose studies have proved that the first appointed traces of plant cultivation in the Old World appeared in the early agricultural Neolithic settlements that developed in the Near East about 7500 – 7000 B.C. All finds till now show that the beginning of food production there was on the basis of 'domestication', i.e. cultivation of comparatively small number (8-9 species) of local cereals. According to Herre and Röhrs, (1977, pp. 245-279). "...the cultivation of plants in the Near East takes place approximately at the same time when the domestication of animals happened, from which the sheep and goats were the first put under man's control soon after followed by other domesticated animals..."

"...Most of the numerous plant remains that were found in the early agricultural settlements were remains of three cereals: einkorn, emmer and barley. The plants are characterized by the wide grains, non-friable ears which are traceable in the archaeological finds come to show that around 7000 B.C. those three annual grass plants had been intentionally sown and harvested in the Near East. ..." Herre and Röhrs (1977, pp. 245-279).

The most often found sowings at that time were the emmer and the barley. According to the archaeobotanical studies the hulled einkorn was to some extent less often sown. From the leguminous the most often found in the early Neolithic Near East were the lentils and the peas. The finds also show two other local leguminous plants - the bitter vetch – *Vicia evilia* and the chick-peas- *Cicer arietinum* Zohary and Hopf (1988).

Nowadays the wild progenitors of the majority of the domestic plants grown in Western Asia, Europe and in the valley of Nile are considered to be for already well identified. 'It is stated that the Neolithic cultivators in Anatolia grew 2 basic types of winter crops – einkorn, emmer and barley. The two-row *Hordeum distichum* L. originated from the two-row wild barley - *Hordeum spontaneum* Koch. It is not clear whether the earliest naked barley is the domesticated two-row or the six-row barley. The wild *Hordeum spontaneum* Koch. progenitor of the cultivated two-row

barley is widely spread in South-western Asia. The basic cereals were the diploid einkorn and the tetraploid emmer. The hexaploid *Triticum aestivum* L., and the *Triticum compactum* Host. were cultivated later. Today it is ascertained that the einkorn is derived from the wild *Triticum boeoticum* Boiss. Some species such as *Triticum boeoticum ssp. aegipoloides* are found in cool regions when the *Triticum boeoticum ssp. chaudar* is found in South-western Livan, Turkey, Iran and Iraq in warmer regions Jakar (1991).

The area of spread of *Tritium dicoccoides* is more restricted and it enters in natural components of the herbaceous formations. It is not found in northern Iran and in the TransCaucasua region but it grows around the western Mediterranean lands and in northern Iraq from where it has started to spread. The emmer prefers colder regions compared to the einkorn. It grows at a height of 750-1000 m above sea level.

The first evidence for domestication of rye came from the Neolithic settlements in the Near East – from Can Hassan III and Tel Abu Hureyra in northern Syria. The finds in Can Hassan III put forward the hypothesis that the rye had reacted to the European soils not as a weed among the other cereals but as an independent plant Hillman (1975, pp.70-73). The Balkans may be had an important role in its migration route.

The bitter vetch is presents in the corn assortment of the Neolithic settlements in eastern Anatolia. The big quantities found prove that it was used also in human consumption – around 7000-6000 B.C. The wild cereals grow in poor and rich soils around rocks and to some extent make the harvest difficult. It is logical to presume that when the cereals were cultivated by the early cultivators, the field should be near to the settlements and in flat terrains Jakar (1991).

Practically, for all early sowings the first evidence of cultivation appeared in the same common areas. The quantity and the quality of the archaeological evidential data significantly vary from region to region. Zohary and Hopf,

(1988) notices that not only the wild wheat and barley distributed in north-western Asia are well known but that similar progenitors could be found also for the rye, the oats, the lentils, the chick-peas, the broad-beans, etc.

Harlan (1987, pp.21-22) on the other side says "that the knowledge about the spread and the ecological conditions of the wild wheat species is sketch".

The cereals are key factor in the examination of the issue of transition from collection to agriculture. The knowledge about the eco - geographical conditions of the wild cereals is of basic importance for the understanding how these plants have spread and about the possibility such plants to enter into the archaeological contexts.

On the basis of series of data from the pollen analysis of the Lake Van, in the Lake District and others according to Jakar (1985, pp. 120-122) it is possible to reconstruct the palaeo vegetation of Anatolia in the Neolithic and the Chalcolithic period – early and middle Holocene. The northern, western and southern parts of Anatolia has covered by continuous forests. The forest-steppe vegetation or steppe vegetation with dispersed forest including vegetation have covered the western part of the central Anatolian plateau, Eskishir plain, Afion province, Lake District. Similary vegetation also covered the southern part of the western Anatolia– Malatya plain, Altinova plain in Elazig, Bingol, Mus, Bitlis regions. Steppe and desert-steppe vegetation has covered the central plateau comprising the Konya basin, Van basin and eastern Anatolia. About 9000 B.C. in some regions of the Near East is noticed intensive evidence of gathering of wild growing cereals. The human populations of the pre-mountainous regions with steppe and forest-steppe vegetation (as with dry and hot climate) started their first attempts at domestication. Bar Jossef and Kislev (1989, pp. 632-642) explain the appearance of the first farming communities and the appearance of the Neolithization in the following way:

— Around 19 000 – 18 000 B.P. small tribes of hunters and gatherers settled alongside the Mediterranean coast and in the western part of the Trans-Jordanian plateau. The cold and dry climate restricted the exploitation of the steppe regions. The special and seasonable spread of the food resources evidences a comparatively mobile style of living.
— Around 14 500 – 12 500 B.P. is noticed a considerable increase in the rainfalls which also increased the exploitation of bigger areas.
— The abrupt climatic change – drying around 12 500 B.P. is the reason for restriction of hunting and the return of the tribes to the Mediterranean coast territories. They increased their adaptation, moved to new territories and thus they turned into semi-settled communities in the Mediterranean region, which later established the basis of the small natuffian permanent settlements.

These three stages could be explained as a combination of climatic changes and increase in the population. The change in the climate between 10500 – 10 000 B.P. and the development of big communities with special organization forced the late-Natufians to adapt to systematic cultivation of cereals and leguminous as well as domestication of animals Josef and Kislev (1989).

Based on the available data from the ensembles of fauna at the end of the Pleistocene the palaeo environment could be presumed in general. The climatic changes noticed in the end of the Pleistocene are less drastic that those notices in Europe. The forest species in Anatolia continued to dominate as regional fauna till the end of 8 000 B.C. Thus for example in Cayönü – South-eastern Turkey – are found deer bones of *Cervus elaphus L.* in 4 of 5 studied layers (Perkins, 1973).

Boss primigenus Boj. is also predominant in the fauna of the same settlement. This species needs open air spaces with dispersed forests. The changes in the natural environment in Anatoilia towards drier climate happened later than in Levant and no earlier than 7000 B.C. The fauna around the Lake District – Suberde shows the presence of wild goat – *Ovis ommon* L. which prefers mountainous places and high plateaus. In addition the fauna in the Lake District – Suberde was of *Cervus elaphus* and *Bos primigenus*. By the end of 7000 B.C. and the beginning of 6000 B.C. the forest spaces around the Anatolian plateau are indicated with the finds of Catal Höyük, which show that *Cervus elaphus* was already much less to be met (Perkins, 1973).

The transition towards the food production in the pre-history called forth adaptation to different economical strategies such as farming, gardening and hending. The epi-Palaeolithic hunter-gatherers have systematically exploited different alimental resources on seasonal niches. The economical system was based on permanent movement and transition from one kind of resources to anothers. The transition from hunters and gatherers to food producers could not be easily traced and proven. The economical evidences in early Holocene are very similar to that in the gathering. It is a fact but that in many regions of South-eastern Anatolia the significance of the nutritious plans increased as proven by many artifacts as millstones. The settled way of living of the farmer communities is based not only on the cultivation and domestication but also on the intensive tillage of land. This wide spectrum involved the gathering of seeds, fruit, nuts and roots. This subsistence economy reflected in the duration of the epi-Palaeolithic system which brought a decrease in the risk of famine at times of reduced crop yields Jakar (1985, pp. 256-258).

Relatively new scientific methods by the reconstruction of the palaeo diet identify strontium or calcium in human skeletons. The presence of strontium and calcium in the human and animal bones as well as in plants could provide a basic indication for the structure of the substance economy in the settled communities with different preferences towards meat or cereal food. The basic change towards mixed farming substance economy in the early Holocene could be due to many reasons and factors. Some of those factors could rely on the environment, others on the economical adaptation by a change in the social organization or to the group competitivenes. At the same time the studies connected with the origin of the early domestication could not be separated from the natural context in which the domestication took place Jakar (1985, pp. 256). According to the opinion of Jakar the fertile plains, the river valleys and the different ecological niches were restricted, which provoked large migration covering huge distance through Greece or the Balkans in 6000 B.C. or some later Jakar (1985, pp. 256).

According to Willcox, (2005, pp. 9-11) the information in most cases is scarce and the findings of many authors Blumer (2002) were not based on direct evidences. Certain differences were reported between the local preferences of the plants and their regional spread (Zohary and Hopf 2000). This spread in the past was connected with the climate shift – by more humid climate their spread increased while by dry one and in arid conditions it withdraw on decreased Hillman (2000, pp. 372–392.). But such factors according to Willcox (2005, pp. 534-541) – as the edaphic or steppe climatic conditions and saltines of soil were not taken into consideration. It is known that the wild cereals and the rye are plants which need calcium and they strictly restricted themselves in such loci. Very often these specific requirements get ignored by the consideration of the origin of the farming. Thus for example the adaptability is the essence of the deterministic model of Bar Josef and Belfer Kohen (2002). The adherents of this model ignore the requirements in relation to their habitat which affects on their adaptation.

The contemporary models also propose the conception for an integrated centre with fast diffusion from it. These models were popular in the early nineties of the XX century but in the last 10 years the achaeobotanical and archaeological data have increased and some of the basic assumptions of these models became questionable. The last evidences favor the polycentric evolution (Gebel 2004, pp. 28-32), according to which the economy was gradually adapted in large spaces. After the labor of land was once established, series of opportunities for domestication appeared in these places (Willcox, 2005).

In that connection the archaeobotanical data of Willcox (2005), Ken-ichi Tanno and Willcox (2006) support this model. They have studied a number of settlements in the Near East dated of 10 000 B.C. and earlier which showed that the ancient cereals have had a model of spread which in general corresponds with their nowadays distribution.

Even by regional differences Willcox (2005, pp. 534-541) supposes that the population made choice of local cereals for cultivation and that it led to independent cases of domestication. These new archaeobotanical data together with the last evidences concerning the domestication of cereals in Cyprus in the middle of 9000 B.C. have not been available where the main hypotheses about the origin of the farming were spread in the nineties of the XX century.

According to Ken-ichi Tanno and Willcox (2006, pp. 197-204) ..."The domestication was a series of processes which happened in different places in the period of thousand years during which time the wild cereals were existing in the cultivated fields (they could be found as weeds even now in Turkey)...". According to Bokonyi ..." the common characteristics of pig and dog was that their food requirements were very similar to those of man and consequently they could easily survive on kitchen remnants. Ideal subjects of domestication were caprivorns also because they could survive a fodder rich in cellulose – straw and other by products of agriculture..." (Bokonyi, 1987).

Thus the data of Willcox, (2005), Ken-ichi Tanno and Willcox (2006) support a model of gradual domestication, proposing the possibility that farming could have appeared soon after the people have accepted a settled way of living in the early settlements of the Near East.

According to Bökönyi, et all., (1973) the sheep and the goats were not domesticated in the South-eastern Europe for the simple reason that the natural habitats of their wild progenitors were in South-eastern Asia. In another words the domesticated species found in South-eastern Europe came from the East where they had already been domesticated. Their herding was ascertained first to the north, north-east and west from Anatolia in the second half of 7000 B.C. This process most probably should have been influenced by the Climatic Optimum in the early Neolithic where the average temperature was with 3-4 degree higher than that of nowadays, which allowed their spread in the southern part of Central and Eastern Europe. The first cattle breeding is proved in Greece in settlement as Argissa-Magula, Achillion in Thessalia, Nea Nikomedia in Macedonia, Lerna and Franchthi cave in Peloponnesus and Knossos in Cryte. There are found animals bones of domesticated animals among which the bones of swines, dog and buffalo. The Neolithic domesticated fauna in Greece seems to have developed there where the local geographical and climatic conditions were similar to those in Anatolia and other regions in South-western Asia Bökönyi, et all., (1973).

Interesting hypotheses about the origin of farming concerning Neolithic Greece is presented by Halstead (1981, 1995, 2000). His study results show intensive farming and breeding in the period 7000 – 4000 B.C. Halstead's arguments are based on pollen data which established the absence of large scale land clearing of the forests. In addition to it Halstead (2000, pp.110-128) states that the Greek tell settlements would have required the maintenance of large herds concludes that because of the availability of irrigation possibilities the growing of cereals and legumes was the bigger part of food sources though the cattle provided alternative vitally important food source in the time of poor yield. The contemporary studies of the pre-historic settlements in Anatolia show that the sheep and goats were reliable enough for domestication because they could survive by feeding of rich in celluloses food originating from the straw that appears a secondary product of the yield.

The recent discussions about the development of farming emphasize on the transition from extensive to intensive form of use of land. The availability of sowing systems is marked with the defined disappearance of the earliest species of cereals in the natural zones and the gradual shift in the sowing cycles differentiating from the wild species. As the evidences for early cultivation and the systems are as different in character as also the resulting from them arguments Sherratt (1980, pp.313-330) proposes 3 aspects which are especially important.

The basic issue in these analyses is the typical system with sowing, burning, very intensive cultivation, leaving the land for a short time and the use of farm technique – both manual and with irrigation. Then other pieces of land get used in the same way. After several years they come back to the previously used fields. By the application of this method the rotation system seems to have been the initial system typical for the early farming groups. This model has had for quite along time a support in Europe where the early Neolithic groups used the land in this way. An additional argument is that this system appears as a distinctive feature especially in the areas with moderate climate, often in regions with broad leaved forests. This is known for the boreal forest spaces as well as for some mountainous regions with coniferous woods in Central Europe and in the Carpathians (Sherratt, 1980).

According to Sherrat (1981, pp. 261-307) a very important and distinctive feature from the study of the pre-historic settlements is their restricted distribution. By comparing the proportions of the landscape with those of today it is established that quite a small part of the land was in use by the early farmers. In the Near East this proportion has been considerably higher in 3000 B.C. when in Europe it grows only in 1000 B.C. and increases in the Middle Ages. Except that the places in use were restricted they were also definitely selective as alluvial, lake and other loci with high underground waters. That is typical for the Neolithic in the Near East and in Europe Sherrat (1981).

But in Anatolia according to Özdugan, (2001, pp. 223-261) the processes are different and they should be considered

separately. Thus for example in Eastern Anatolia the Neolithic development is more progressively, filling different natural niches which is connected with the population density while in Central Anatolia the Neolithic develops more gradually. Except that the Neolithic here outlines definite boundaries with Neolithic zones. To the north and to the west of these zones the things change and thus Central Anatolia represents a way of boundary with the other independent interior zones of neolithization.

Özdugan states that in Western Anatolia – Cayonu, Nevali Cori, Halan Çemi and others on the basis of the architectural details and others these settlements are not used as ordinary living premises but they have had restricted use. For that reason it calls them "Temple", or special buildings. Evidence for it are the found objects, idols, etc. which are in quite a higher concentration in these sites. As result of it he proposes the central Anatolia to be considered for a boundary, a bridge presenting different models of neolithization Özdugan (2001).

In the early phases of farming the settlements of special significance such as Jericho and Myreybut stay outside of the zone of farming without irrigation. According to data from the tell settlement Myreybut (10 450 – 9 950 B.C.) there are the earliest archaeobotanical finds – of wild einkorn and rye. Similar species are found also in Aby Hureyra, where the early cultivation could have come into being. The results show semi-seasonal occupation and namely minimal occupation in summer and spring but for the late summer and winter there is no confirmation. These data are stated by Willcox, (2005, 2006). According to him it seems that the settlement has had a seasonal character of inhabitance.

The end of the Glacial in Anatolia the early Neolithic settlements and more specially those in the Konya plain involving Catal Hoyuk were situated in regions where the rivers run over arid plains. The early settlements in Greece and in the Balkans are concentrated in regions with tectonic basins, near to springs or large river terraces with seasonal floods Sherratt (1980, pp. 313-330).

In Central Europe the early Neolithic settlements occupy the Middle Danube plain and places by rivers and springs with loess soils. Contrary to this model of "shifting cultivation" there are enough settlements characterized by long period of occupation lasting several hundreds till thousands of years. That is typical for the culture Bandkeramik i.e. for their settlements in North-western and central Europe as well as for the tell settlements by the rivers in the Near East. The early farmers have occupied only small zones characterized by maximum productivity of the local systems of cultivation. Such zones were provided with running water that favored the growing of cereals and the fishing Sherratt (1980, 1981).

Similar is the concept of Bogard (2004), who propose that in the Neolithic period the sowings were located predominantly in small intensive fields. Bigger fields appeared only in the Middle Ages. To some extent that is proved by the ways of harvesting techniques.

The presence of weeds buy harvest or after it depends on the methods of harvesting. For example in 'high' harvest done by sickle the cereal ears and the weeds which are at the same height as the wheat are to be taken together. And by a 'low' harvest are they collect not only those weeds which have reached at the same height as the cereals but also such ones that grow lower as the clover. The finds of low growing weeds in the Late Bronze Age and in the Early Iron Age in the territory of Europe prove that the harvest has been a low one whilst in the Neolithic the 'high' harvest done by sickle was practiced – evidenced by finds characteristic for high harvest. She ascertained also that in the Neolithic fields were found more perennial weeds while much later in the Middle Ages the annual ones already appeared which is connected with the permanent cultivation of large areas (Bogard, 2004).

By the end of 4000 B. C, and the beginning of 3000 B. C. the use of the plough became wide spread and thus the expansion of the arable lands become inevitable, the forests thinning more intensive and the breeding significance grew.

According to Sherratt (1981, pp.261-307) a characteristic feature of the early farming communities in the Near East is the special decrease and restriction in the alluvial zones. The soil humidity was extremely important as their farming systems relied on underground waters as well as of great importance were also the seasonal rainfalls. And the wild cereals are strongly adapted to seasonal changes. The wild corns and barleys are representatives of the opened plant communities developing mainly on limestone or basalt lands. They are found on plateaus in the sub-Mediterranean forest spaces. The development of the first stages of cultivation is connected with the laboring of the soil and its systematic rotation. Such tillage is characteristic for the thick brown soils in most of the steppes in northern Assyria in 6000 and 5000 B. C. This model could be supported by mattock farming by laboring the land from 1 to 4 years. Gradually the sowings moved from zones with underground waters to zones with shallow waters.

The additional laboring developed with the appearance of the new techniques – such as the plough that appeared about 4000 B.C. in Mesopotamia. By it the soil gets friable and its capillary activity increases.

In Europe but this technique is less widely spread as here it is necessary first to do forest thinning Sherratt (1981, pp.261-307).

The cultivation in the inter river valleys requires winter seasonal sowing while in the moderate zones of Europe the high rainfalls allow the duration of the spring sowing. In the late Neolithic and in the Bronze Age the main place kept the barley sown in the spring as its cultivation was favored by the light soils.

The first evidence of winter sowings in Northern Europe is the spread of the spelt - *Triticum spelta* – in the first millennium. It seems that at that time the winter varieties of the other cereal species were also defined as productive. The spring sowings adapted where they could survive in the northern regions, i.e. where the winter is cold and lasts longer Sherratt (1981).

THEORIES FOR RECONSTRUCTION OF FARMING AND ETHNO-MODELS

Today, for the reconstruction of the farming different contemporary approaches for interpretation of the archaeological data are used. One of them, considered with special attention by Bogard, is the information about the weeds.

From the archaeobotanical studied carried out in the Deliorman region in South-central Romania Bogard (2004a, pp. 49-58) establishes small groups of potential 'arable' weeds found in 'flat country; settlements in Kric, Dubecti – culture Boyan (VI-th - early V-th millennium B.C.). The restricted evidence material found till recently shows intensive farming with development of weeds often similar to those from the Neolithic settlements in Central Europe.

Another source of information for reconstruction of the agriculture comes from the phyto-sociology which contributes to the interpretation of the appearance of characteristic species in the archaeobotanical remains as indicators of the living conditions. Some authors as Behre and Jacomet (1991), Küster (1991), Gones (1992) think that "...with certain confidence most common groupings of species could be applied to the archaeological weed groupings as well as the appearance of characteristic species which belong to these common groupings could be used as indicators for the living conditions by which this group as a whole did appear..."

Another essential problem according to Charles, et al., (1997) exists for example by plant communities which are observed only during their growth but no differentiation are done between their ecological requirements and their tolerances. Thus species in one plant community which germinate and grow by condition of humidity could be considered for indicative of dampness if even only some of them tolerate certain humidity but have their other specific requirements as for example soil fertility etc. Or, presented in another way, the terrain observations connected in the phyto-sociological communities with the conditions of the vegetation growth does not provide enough data which exactly aspects of the environment have become the reason some species to grow in definite places.

Lately the phyto-sociologic data are widely used in the archaeobotany to arrive at conclusions about the environmental conditions and for practices of the sowing farming van Zeist, (1974), Willerding (1979, 1983a), Jacomet et all., (1989, pp. 128 - 144).

Chapter 7

ECONOMY

CEREALS

SOWINGS – ethno-models

To establishing the origin of the plant remains we need to study them taphonomically – way of conservation and methods of grain processing. The archaeobotanical material contains quite a lot of grain remains as well as seeds, weed, glumes, straw and chaff. In that connection the ethno-model of Hillman and Davies (1992) characterizes the effect of the harvest stages and their connection with the composition of the samples in the archaeobotanical contexts (on the basis of the ethnographic models). They provide a very simplified scheme of the different types of activities:

— Harvest – harvesting the cereals
— Threshing – dismissal of glumes, straw and chaff from the corn
— Winnowing – winnowing of the light glumes and ears and stem fragments and of the light weeds seeds
— Sieving through a coarse sieve – elimination of the heavy big sized weeds and the fragile stems
— Sieving through a fine sieve – winnowing and clearance of the small weeds from the grains Hillman and Davies (1992, pp. 113-158)

Thus after the clearance of the grains the harvested cereal crop goes through different processes each of which presents a basic product or a secondary product which could be identified as different proportions of grains, chaff, straw and weeds. Many of those products and by-product have short presence or get mixed with other secondary products. Some participate longer and they are exactly those which could be found in the archaeological contexts. Such products usually derive from the following activities:

— Winnowing of a secondary product
— Coarse sieving of a secondary product
— Fine sieving of a secondary product
— Fine sieving of a product.

In conclusion it could be said that in each of those processes we could have different ensembles. And each stage creates a product and secondary product some of which could get preserved in the archaeological contexts Hilmann, (1981), Hillman & Davies (1992).

By the study and reporting of data it is important to know that the different cereal components have different conservation and it exerts serious influence on the interpretation. The experimental data show that the spikes of the naked cereals are ones of the first components that burn immediately, often burn also the glum and the basis of the spike burn also. The situation with the hulled cereals is more complicated. By examining the remains of the stem the cereals could not be interpreted because it is extremely difficult to be defined and, except it the straw could have been moved from another place brought as fodder for the cattle.

Still it is possible to define the secondary products in the early stages of two wheat species which contain different weeds. But when the samples contain both wheat and barley, the weeds are to be such as they are characteristic for the both cereal types Nesbitt and Samuel (1995, pp.41-99).

Studies connected with the archaeological contexts as well as such based on ethnographic models are done by Palmer who studied a certain region in Jordan.

In northern Jordan both winter and summer sowings are practiced. The main plants are grown: durum wheat - *Triticum durum* Desf, *Hordeum sativum* L. - hulled two-row and six-row barley, *Lens culinaris* Medik.- lentils, *Vicia ervilia* Willd.- bitter vetch. These are the basic winter plant cultures but other winter cultures include also the broad bean – *Vicia faba* and the common vetch - *Vicia sativa*.

In mostyof the studied by Palmer, 1998 sites, the fields have rotation regime and enough humidity. A two or three year rotation regime was practiced. It is established that two-year rotation regime is practiced on the slopes and three-year rotation regime is used in the valleys and the plains. The two –year regime on the slopes is called winter-summer regime and the winter plants usually are wheat followed by summer sowings. By the three-year regime the wheat is followed by leguminous and then again by winter sowings.

Similarly ethnographic data from Bulgaria here provide the following interesting information:

"…Usually for fields flat lands were used. Where they were hills and mountainous lands smaller pieces of lands were used. The soil gradually went down to the lower ends of the fields and thus in the course of the time to some extent 'natural' terraces were formed. By more arduous lands mostly in the Rhodopes and alongside Struma River artificial terracing was done. The preparation of the field was made as initially the clear land. The sows were left to burn through the summer after which the field got crossed reploughed and in some places dragged by a harrow to make the clods smaller. The places got cleared by stones which were heaped up by the ends of the field to serve as fencing. Comparatively more complex is the processing in the forest lands – the simplest way in such a case is the woods to be cleared at place. Then the new stems that sprang up were each year consistently knocked off till their roots got dry and then in the order of their decay the roots were taken out. When there were bigger trees whose uprooting was difficult they were dried through barking. This method was widely used in Southern Bulgaria, mainly in Strandja and Sakar. The created in such a way fields were called 'knocked-off' ('chukanini') – in the districts of Pazardjik, Elhovo, Samokov. Another method was the burning of the forest. Usually after the fire on the ashes was sown rye. The sowing of such a field took place after the spring. In the next several years the field was sown again but already prepared by tillage or ploughing. And thus till the soil get fully exhausted. The new fields were sown first with spring sowings – rye, barley, oats and millet. The winter cereal crops were sown after the soil was better used. The first plough in the spring was called 'turning' while by the second one which took place in the summer the field got crossed ploughed. By fields with stronger soil the sowing system is of two-years. By it there is no rest of a whole year – one year is sown a spring species and the next year – a winter one" Vakarelski (1974, pp. 108-111)

Ethnographic models connected with cereal crop processing and the different phases of this activity are developed also by Hillmann (1981, 1984).

Thus for example there exist different debates on the methods of harvesting – 'high; harvest (only of ears) or 'low' harvest – together with the stems. The harvest always begins from the 'back' of the field, i.e. - from the side from which the wind has bent the wheat. Otherwise it is not possible to work against the ears. The separation of the grains

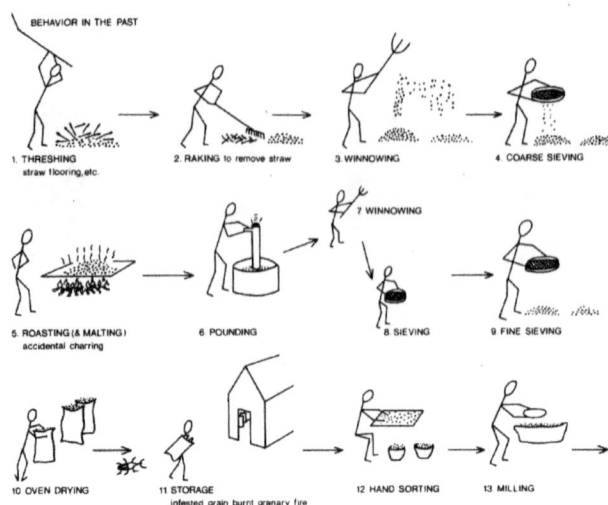

Fig. 26. Exemplary scheme of the processes presented by Renfrew & Bahn –different stages of the cereals crop processing.

from the ears is done in several traditional ways – from the simplest manual knocking-off to cattle threshing. Except by the cereals by the leguminous after its transportation the bitter vetch also gets threshed with threshing-board and the separated grain is used as food for the cattle and the straw is food for the sheep (fig. 28).

The ethnographic data in our lands indicate than in the Middle Rhodopes where the sheaves are smaller they get knocked off being hold in hands and their ears hit on a stone or wood put on the ground. The regular threshing but takes place in an opened air place. This place gets cleaned very well and trampled on. The place for threshing is called 'harman' (stack-yard, threshing-floor). On it the sheaves are arranged concentrically each by the other with their ears on the previous ones. The cattle are bound by a long rope attached to a fixed wooden pillar in the centre to make it go around trampling on the sheaves and the rope gradually coils around the pillar till the cattle reach to the middle of the circle. Other than using cattle especially in Eastern Bulgaria, the threshing is done by threshing board Vakarelski (1974, pp. 10112-117).

Storage

The ways of storing and cleaning of grain is a controversial question.

Many facts prove that the cereals in the pre-historic times were stored as ears. Thus for example in the site of Yunatsite - one of the samples was taken from a granary (dated Early Bronze Age) where the quantity of the find of barley was about 50 kg. What made the impression was that big part of the find consisted of whole ears. It provides evidence that in this case the concerned find is of a big storage and of rich yield. Except it in this context we receive also valuable information about the manner of harvesting techniques – and namely if a 'high' or a 'low' harvest was practiced. In the case the harvest was a high one – the stems were not gathered but only the ears.

Hilmann (1981) presumes that by humid conditions as those in Northern Europe or in mountainous lands after the harvest the whole yield of grain should needed to be cleared whilst in dry climate the cleaning of the grain could wait for some time. In such a case small quantities of grain get cleaned – as needed for the day, day by day.

Usually the cereals were stored together with the glumes. Such small quantity of cereal was found also in the site of the Madretz, near Nova Zagora. The grains were established whole ears with glumes and also separated glumes and cleared grains of einkorn, emmer and spelt. Obviously the grains were collected and already in the process of cereal crop processing.

Not only were big pithoses used for storage but often vessels were also used. Such are the found vessels – in the site Suvorovo.

An interesting example of way of storage is provided by the finds from Hotnitza where there were found different mixtures of cereals in several heaps. According to the data provided by the archaeologist Chohadjiev and Chohadjiev (2005, pp. 9-12) these were pouches made of cloth which later on burnt.

Ones of the basic ways of storing is the subterranean pits and pithoses. According to Sigaut (1988, pp. 3-32) the subterranean pits are the standard milieu for hermetic storage in many parts of Europe, the Danube plain, the Balkans and in Ukraine. In some of the studied settlements there exists direct evidence for usage of pithoses and pits.

Charred plant remains were found in a pithos from I building horizon in Vaxevo settlement. It was put in a pit. The content of the cereal mixture is about 150 grams and it consists of cleaned grains of hulled barley, emmer, einkorn, rye, common/durum wheat. Similar storages are found in pits (pit No 3 in sq. H 12, pit No 4 in sq. J 4) by the site Adata.

The found cereal remains in the site of Yabalkovo – from sq. I 39 – in an early Neolithic pit No 2, could be also a kind of storage.

On the other hand it is quite different to identified if the stored crop was intended for man or animal use. The storage conditions of the products could not prove their usage. Nor the identification of the food by basic types could provide evidence for it. By presumption the common wheat and the peas were grown for the people while the oats and the bitter vetch could have been intended as animal fodder.

Fig. 27. Contenmporary threshing board. Center for Anatolian Ethnography and Textile Studies in Istanbul, Josephine Powell, 1999

Experimental data in relation to the yield were achieved by a number of authors with different variability.

The yield of grains is measured in small unites and usually as a proportion of the quantity of grains for sowing and the quantity of grains achieved by the harvest. Such a correlation is usually 1 : 2 or 1 : 3 by poor yield and 1 : 8 and 1 : 10 in good years by rich yield Sigaut (1992, pp. 395-403).

According to Jakar (1985, pp. 276-277) the annual yield of the wheat and barley in Neolithic Anatolia depends on a number of factors such as climate altitude, hydrology, soil type and farming technique. He presumes that if an average the yield is 200 kg per dca and that could be accepted as an average yield from regularly watered land, then a minimum of 2000 calories per day are necessary for a man, which makes approximately about 700 grams of bread and pulses per person per day. And, if the one husbandry consists of 5 people there are necessarily 5 kg bread per day for it. In such a situation the family should farm from 6 to 8 dca of land. And here should be mentioned separately the additional sources of food from the vicinity.

The grains yield by the traditional techniques in the Mediterranean is about 300-600 kg per hectare. Halstead (1981, pp. 317-318) presumes that by a more intensive practice in the Neolithic the yield could have reached about 1000 kg per hectare.

Bakels (1978) calculated that for a minimal area in the early Neolithic period 50 people were needed (LBK in central Europe – 5400-4900 B.C.). She based her calculations on ethnographic models from Russia and Canada as well as on experimental data. Thus she established than no less than 11 ha were needed to 50 people for sowing cereals and achieving a yield of 800 kg/ha if their food consists

65% of cereals. In addition there 150 ha more were needed for the cattle.

Nowadays in the region of Krushare, Chokoba - Sliven district 200 kga wheat or about 120 kg barley are achieved from 1 dca. For the feeding of a single goat a about 100 kg grains are needed and additionally some 5 packs of lucerne and 5 packs of straw (local information).

Hilmann and Davies (1992), complement the data about the quantity of corn needed for a family consisting of 5 persons. Their calculation is based on the caloric consistence in contemporary wheat so they presume that 255 of all needed calories would be achieved solely through the corn. In such a case for a family of 5 330 kg. per year would be needed. And if the yield was of 500 kg/ha per year then for it 0, 75 ha should be sown. And if the yield is rich as 1000 kg/ha then there will be necessary 0, 4 sown for 5 persons per year. On the basis of the data from the ethno-models in Anatolia. Hillman and Davis (1992) affirm that 330 kg of wheat are needed for a family of 5 persons annually whereas for a single person 114 kg should be needed.

Based on ethno-historical data Borochevic, 2006 also reports that by sowing 1,5 ha around 900 – 1200 kg corn are produced for which about 1000 hours human work in the field are needed – 600 hours for manual sowing, 20 hours for sowing by throwing seeds, 300 hours for cutting with knives made of flint and 60 hours for knocking-off. If animals are used for ploughing then are only 30-40 hours are needed. And if the working time is of about 9 hours daily then a man needs to work 4 months for a field of 1,5 ha. And if more people are working then 1, 2 months would be needed. This calculation is but only for the wheat sowing.

Except for the preparation of bread should be noted that from cereals were produced also gruel and grits. Pounded up grains of wheat and barley have been evidenced in some archaeological sites Gross, et all., (1990). The preparation of grits and similar finds were found also in the territory of Bulgaria. A consistence of the kind of a mess was found in the site of Madrets – the grains are pounded up and deformed Popova (1995) as well as in the site of Kapitan Dimitrievo Marinova (2001).

In Mesimiriani – a settlement dated from the Bronze Age – pounded up charred wheat was found, Valamoti (2004) presumed that it was meant for grits.

Traces of straw are often found in the studied negative contexts as pits. They present stems, glumes and parts of the ears of the grains of cereals which show that those pits or granaries were burnt and later used over again. A part of such mixture often shows presence of glumes remains from einkorn and emmer.

The straw has been a basic food product in winters in cattle breeding. As a primary food is the straw of barley, wheat, bitter vetch and the lentils. One of the important roles of straw and chaff is that they appear as basic component in the clay used for elaboration of ceramic vessels, in wall construction, etc. It is like this as on one side it makes the constructions lighter and on the other side the cereals stems contain siliceous elements which are a precondition for the strength of the cereal stem.

Such an illustrative example of roof construction elaborated by einkorn straw is found fragments felted with clay from the settlement of Hotnitza. Small branches were very attentively separated from it – they were the stems with their spikes. The separated material provided the opportunity the stems to be defined as stems of einkorn.

As a rule the houses were built using fences finishing with grout.

Some contemporary ethnographic data of Palmer (1988, pp.1-10) from Jordan prove how the straw of barley and wheat was used. The wheat straw is call-ed 'white straw' while that of leguminous plants is called 'red straw'. The two different types of straw differ also as raw, hard and soft. The soft straw and the lighter straw get separated by the winnowing and they are used as sheep and goats food. The hard straw and principally the harder components are waste from the coarse sieving. Usually they are used as fodder for the horses and the donkeys. The finest straw, the very light one goes to the poultry.

There are certain differences in what the animals eat – the leguminous cultures are more important for the sheep, the goats and the cattle as well as broken grains. The red chaff is more nutritious than the white one and it is more suitable for the sheep and goats than for the cattle. The better quantity of the red chaff is confirmed by the local farmers as also by experimental data. The study of protein contained in the leguminous chaff such as the one from chick-peas, lentils and broad-beans shows significantly higher values than those found by the wheat and barley. The chaff from bitter vetch is highly praised even more than that lentils. Most probably the high percentage of the bitter vetch in the Neolithic settlements is due to the higher usage not only of its grains but also of its chaff.

The most important plants found in the studied pre-historical settlements in the territory of Bulgaria are the einkorn and the emmer. In many of the contexts they are together in different proportions. The cereals could be both sown in winter and also in spring.

Barker and Gamble (1985) suppose after the migration of the farming in the Balkans around 6000 B.C. that the traditional Mediterranean farming in its early stages includes the cultivation of autumn sowings and a progress of the

Fig. 28. Manual threshing of the grains. Center for Anatolian Ethnography and Textile Studies in Istanbul, Josephine Powell, 1999.

Fig. 29. Threshing of the grains with threshing board. Center for Anatolian Ethnography and Textile Studies in Istanbul, Josephine Powell, 1999.

Fig. 30. Winnowing of the glume from the corn. Center for Anatolian Ethnography and Textile Studies in Istanbul, Josephine Powell, 1999.

animal husbandry while thousands of years laterly different other species based on spring sowings get included.

In many Balkan settlements including those in Bulgaria the einkorn and emmer are found together and probably they were sown together Hopf (1974). Such mixtures called by Jones and Halstead (1995) – wheat mixture "maslins" could have been grown to reduce the risk of loss of one of those two components. But I think that there is still not enough reliable evidence for it as also in many places in the settlements studied by myself the two have been established separately.

LEGUMINOUS PLANTS

The palaeobotanic finds show a diversity of cultural plants among which the leguminous are also represented. The available archaeological evidence proves that the lentils, the peas, the chick-peas, the grass pea have been cultivated more or less together with the basic cereals. Their remains are dated in numerous Neolithic settlements in the Near East. They are also often found in the context of the Neolithic sites which appeared some time later all over the vast territory from the Atlantic coastline of Europe to the Indian sub-continent Zohary and Hopf (1988).

The basic cereals and leguminous plants in the Balkan Peninsula lands have appeared at the beginning of the Neolithic. The first evidences refer to Palaeolithic and Mesolithic strata of the cave Francithi Hansen (1978).

In Neolithic settlements in the territory of the country sustainably settled several basic cereal species – einkorn and emmer but together with them arrived also the leguminous plants. They have been domesticated at the same time as the cereals. In that connection are scrutinized some of their basic species which have played a substantial part on an equal level with the cereals in the feeding of the local population.

Vicia ervillia Willd. – Bitter vetch

Bitter vetch – A representative of the leguminous plants the bitter vetch is documented in almost all studied settlements from the Neolithic to the Bronze Age. It is traced permanently through all these periods. Quantitatively in some finds it is presented only as single grains in Golyamo Delchevo, Omurtag, Slatino, etc. while in others it is found in big quantities probably as a storage – Hotnitza, Varhari, Drinovo, Podgoritza, Galabovo, Azmak, Nova Zagora, Madretz.

Different quantities of this species are reported by Marinova in the settlements she studied – Kremenik, Galabnik, Kapitan Dimitrievo, Karanovo, etc., which refer to early – late Neolithic Marinova (2002).

The percentage correlation of the bitter vetch by epochs and by settlements is presented in fig. 31 where it can be seen that still in the Neolithic the bitter vetch participated with a high percentage.

The bitter vetch is one of the oldest plants which apparently have been initially domesticated in the territory of the Near East - Northern Iraq and Iran and especially in Asia Minor Zeist van (1972), Knorzer (1973).

From the studied settlements presented in fig. 31 it is stated that the bitter vetch has the most wide spread in the Neolithic.

In the territory of Greece for last time bitter vetch is documented in finds referring to the late Neolithic in Sitagroi, Dikili-Tash Valamoti (2004).

It is obvious that the bitter vetch is the more often found leguminous plant in comparison to peas and lentils. Apparently it was more preferred but its toxicity is known so that its seeds have to be soaked in water before their consumption. On the other hand the plant is quite unpretentious towards humidity, and it endures daught and that should have been an advantage over the other leguminous plants which still require some irrigation.

The bitter vetch continues to be grown later in ancient Greece and in Italy.

Theofrast calls it 'bitter vica' and specifies that there are several species. Plinius mentioned that it was used for flour from which bread yeast is prepared but its main usage is as fodder for the ox. In a later stage the bitter vetch plays a secondary part as animal fodder. Today it is sown in very restricted quantities in some countries of Asia. Probably it was initially used more in human nutrition but gradually its role decreased and it turned into a secondary product for animal feeding.

Interesting information about it comes from local inhabitants in neighborhoods in the Eastern Rhodopes nowadays who report that they use it as finely ground supplement to their coffee drink, which proves one more unknown and quite interesting way of usage. Thus confirms that it still grown in some regions in the country.

Lens culinaris – lentils

Lentils – its seeds found in almost all studied sites are of the same type irrespective if they belong to earlier or later periods. The average diameter of the lentils seed is 3 mm, which defines it as the species *Lens culinaris var. microsperma*. It originates from the Near East and, having in mind the available materials it appeared much earlier than the cultivated lentils - *Lens culinaris var. macrosperma*, which originates from the Mediterranean regions. That is proved also by the lentils seeds found in Jeriho, Djarmo, Tel Saabs, Ali-Koch – all those settlements are located in the territory of the Near East Renfrew (1973)., i.e. the lentils finds in the territory of Bulgaria are part of the entire Anatolian complex. From the data we have at our disposal its presence in the studied periods could be traced. It could be supposed that the increase in most of the settlements referred to the Bronze Age is connected with better developed farming and much more successful irrigation technique.

The cultivation of lentils is connected with that of the cereals and the barley in the Near East. Most probably this legume had been cultivated in the same region together with the emmer, einkorn and the barley which should be considered as basic cultures in then Neolithic farming of the Old World Zohary and Hopf (1994).

Small charred seeds of lentils were found in Mureybut (9200 - 7500 B.C.), van Zeist (1970) and in the tell settlement in Abu Hureyra Hillman (1975), in northern Syria. In these early Neolithic settlements the gathering of wild growing lentils *L. orientalis* is witnessed together with the gathering of wild einkorn and wild barley. It is presented

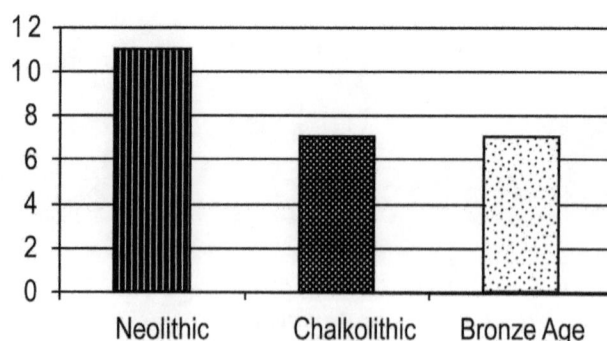

Fig. 31. Quantitative distribution of the bitter vetch in the studied settlements.

in most of the early (7-th millennium B.C.) settlements in the Near East: Ali- Kosh, Helbaek (1969), Iran; aceramic Haçilar Helbaek (1970); Çan Hasan, Anatolia Renfrew (1968); Tell Aswad van Zeist and Bakker – Heeres, (1979); Abu Hureyra, Syria Hillman (1975), and in the aceramic Neolithic Jericho.

Analyzing the remains of lentils from the early Near East regions it is difficult to decide if they are wild growing or cultivated species because the lentils seeds are very similar in the wild and in the cultural species. Only the increase in the size is an indicator. The process of increase of the seed sizes is gradual and the first lentils seeds with some bigger size appeared about the end of the 6000 B.C., i.e. around 1500 years after the permanent settlement of the cereals in the Near East.

In the territory of the Balkan Peninsula several small sized seeds for palaeo and Mesolithic strata were found in the cave Franchthi in Greece. The seeds are small: 2, 5 -3, 0 mm Hansen (1978, pp. 39-46). The archaeobotanical finds reveal that in the 6-th millennium B.C. the spread of the lentils could be closely connected with the spread of the Neolithic farming in South-eastern Europe where lentils remains were found together with of cultivated emmer, einkorn and barley.

Remains of lentils are available in almost all Greek early Neolithic settlements as for example the aceramic stratum in the site Gediki.

Renfrew (1979), the aceramic levels in Argissa – Magula Hopf (1962), Sesklo, Kroll (1981), in Knossos - Greece Renfrew (1979). Lentils finds are evidenced also in many archaeological sites in the territory of Bulgaria and in the countries of former Yugoslavia. Remains of lentils plants are found in Karanovo Hopf (1973) and in the culture Anza - Starchevo – from 5300 - 4 500 B.C. in former Yugoslavia Renfrew (1976).

Numerous finds of lentils come from the studied settlements in the territory of the country. Lentils are found in the Neolithic sites: Durankulak, Vratitsa, Yabalkovo, Drenkovo – Ploshteko, Eleshnitsa, Kovachevo, Koprivetz, Malak Preslavetz, Samovodene, Slatina, etc. and they are traceable in all other periods of the pre-history (fig. 32).

Pisum sativum – peas

Peas – in quantitative relation this species is very often documented as a single grains. Fig. 33 shows that it participates in all studied periods. As a cleaned plant culture and in a significant quantity it is found only in the settlement Madretz – culture Karanovo II. Its average size is 3, 3 mm. By this feature the find there refers to *Pisum elatius*. In most of the settlements the peas have similar morphological characteristics. Analogical are also the data from Sadovetz defined at the time by Prof. Arnaudov as *Pisum elatius*. Dennell (1976) also noted finds of peas in Ezero, Chavdar and Kazanlak but no quantitative data were mentioned. Finds of peas for sowing are defined by Marinova in the settlements: Durankulak, Karanovo III, Galabnik, Kremenic, Slatino, Kovachevo, Kapitan Dimitrievo where in some of these settlements it is presented by single grains while in others it is in bigger quantities.

Considering my data and those of Marinova (2002) the peas were less widely spread compared to the bitter vetch and it was grown in more restricted scope which most probably was connected with its mesophyly as the climate in the Balkan peninsula was not so favorable for it as was the moderate climate in northern regions of Europe. As evidence for this finding is the fact that in part of the studied settlements in the north-eastern Bulgaria the quantity of the found peas is higher than in some other regions and that could be explained by the cooler climatic conditions there.

Peas sowings were presenting also in the early Neolithic farming settlements of the Near East (7500 – 6000 B.C.). Charred peas grains are found in the aceramic strata of Jarmo in Northern Iraq Helbaek (1959); in Çayönü- Southeastern Turkey van Zeist (1972); in Tell Assad, Sothern Syria (van Zeist and Bakker – Heeres (1979) and in the Neolithic aceramic layer in Jericho Hopf (1983). Numerous much richer finds of peas remains are available from some Neolithic phases in the Near East – from the 6-th millennium B.C., such as the finds in Çatal Höyük (5850-5600 B.C.); Haçilar (5800-5400 B.C.), Helbaek (1964, 1970). Most of the observed seeds have significant variations in their sizes. The palaeobotanical studies of Waines (2007) of materials from Khirokitia provide information about the availability of peas, also small sized – 3, 0 0 - 5, 5 mm, which the author refers to a species close to that found in Jerihon B. Similar results come from the seeds found in Titris Höyük – Early Bronze Age– Northern –east Anatolia Algaze, et all., (1995, 2001).

The measurements of the pea seeds found in the territory

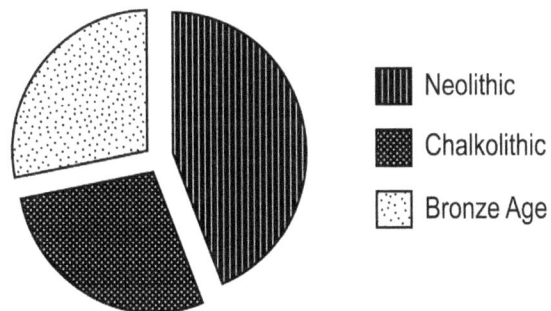

Fig. 32. The percentage correlation of the lentils during the different periods.

of Bulgaria come near to those found in Asia Minor.

Its representative early locations in Greece include the early Neolithic Nia Nikomedia, around 5 500 B.C. van Zeist and Bottema (1971) and the aceramic strata in the archaeological sites Gediki, Sessklo и Sufli Renfrew (1966), Kroll (1981). Charred peas grains are found also in the late Neolithic contexts in Gomolava van Zeist (1975) and Valaç Hopf (1974), in the lands of former Yugoslavia.

Peas are represented in early Neolithic sites in Bulgaria. The finds of Azmak tell are dated of about 4 330 B.C. Hopf (1973).

The contemporary data refer to several sites studied by the author and by Marinova: the Neolithic – Yabalkovo, Drenkovo – Ploshteko, Eleshnitsa, Kapitan Dimitrievo, Koprivetz, Ovcharovo, Hotnitza, etc. Leschtakov, et all., (2007), Marinova (2002).

Cicer arietinum L.– chick-peas

Thanks to some new archaeobotanical studies of significant interest are the first time finds of chick-peas. They are from the Chalcolithic site near to the village of Orlitza, Kardjali district, from Chalcolithic layers of the tell settlement by the village of Yunatzite and Hotnitza, and from the tell settlement Kapitan Dimitrievo – which proves that the chick-peas were cultivated together with the other plants that arrived from the Mediterranean lands Marinova and Popova (2008).

The earliest finds of chick-peas are found in Northern-western Syria – Tell el Kerkh Willcox (2006). In Turkey and more especially in Northern-western Anatolia the chick-peas are found in several early Neolithic settlements – (Catal Höyük -8240-7760 uncal.B.P.), Fairbain, et al., (2002) and in the Chalcolithic site Kurucay Nesbit and Samuel (1996).

The Neolithic finds from South-eastern Europe refer to Greece - Otzaki and Dimini Kroll (1979, 1981). The plant was reported by Kroll в Thessaly Kroll, (1991).

The chick-pea is highly appreciated in traditional farming in the lands of the Mediterranean basin and in Western Asia. Its area is more restricted than that of some other leguminous plants – the chick peas is adapted to the warm Mediterranean climate and it does not tend to develop successfully in regions with cooler climate.

Lathyrus sativa/ cicera – grass pea

The grass pea develops in dry lands and arid soils. Most probably these features of the plant attracted the man's attention in the antiquity for starting to use it and to grow it intentionally. Nowadays this cultivated plant is mostly used as animal fodder.

The comparative archaeobotanical studies of the cultivated plant grass pea in the past epochs and today come to show a close morphological similarity in the plant remains from archaeological sites of this plant with those of a group of wild growing species *Lathyrus*, which are distributed alongside the Mediterranean basin and in Southwestern Asia. The contemporary studies prove that this plant is closest to *L. cicera* – wild growing plant which could be found in some regions of the Eastern Mediterranean and in the countries of the Near East (in Greece, Turkey, northern Iraq, Northern Iran). With its migration to the west the plant adopted qualities which are characteristic for its cultivated species. Such species are found in Algeria, in Spain, in South France and in Italy – and these countries appear exactly the regions of spread of the earliest big grained species Sinskaja (1969).

Charred seeds of the grass pea are found in several Neolithic settlements in the Near East, in the Aegean region and in the Western Mediterranean. They are ascertained among the other finds from the 7-th millennium B.C. from the site Djarmo Helbaek (1960), found are also in the late Neolithic site Dimini in Greece Kroll (1979), and from the Azmak tell - 5-th millennium B.C. Hopf (1973).

The contemporary data show its wide spread in the Neo-

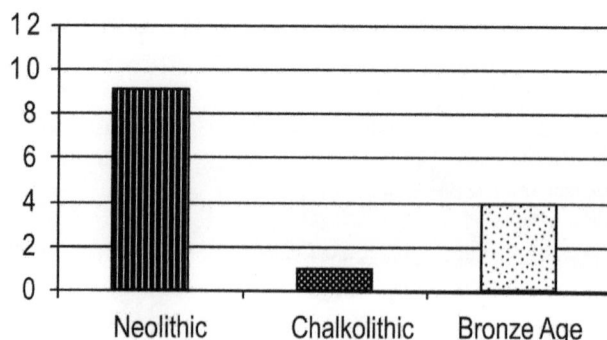

Fig. 33. The percentage correlation of peas during the different periods.

lithic in the territory of the country where quantitatively it is more often found in the Chalcolithic settlements.

The data from materials of leguminous plants in the studied sites are presented in fig. 34. From these findings it could be established that in the Neolithic and in the Chalcolithic the bitter vetch is dominant whilst in the Bronze Age the usage of the lentils increases. The peas should have been less preferred which, as it was mentioned above, is connected with its mesophyly.

Examining Bulgaria as part of the region of the Balkan Peninsula, the finds in each of the neighboring countries could be used for filling the gaps of knowledge for a given period of time in the regional development of the agriculture.

3. ARCHEOPHYTES (WEEDS)

More information about the cultivation of the cultural plants is obtained through studying the weeds. The weeds characterize the environmental conditions, the type of soil, climate and the specific agricultural practices – manuring, sowing, etc. "ARCHAEOPHYTE" is called– weed, each plant that inhabits is the arable land. Thus for example the *Agrostema githago* (corn-cockle), the corn-thistle,

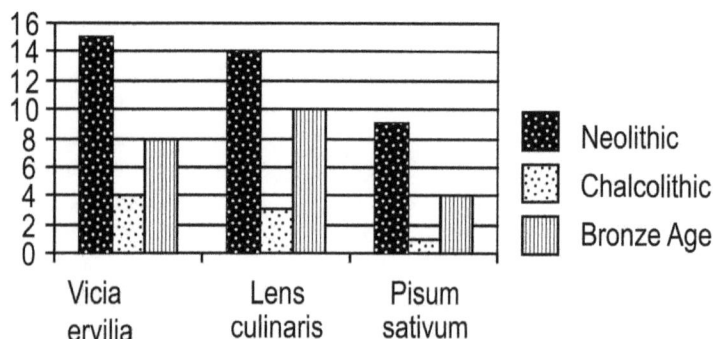

Fig. 34. The percentage correlation of the different leguminous cultures during the different periods.

etc. are archaeophytes. Wild growing plants are connected with man activity. Here are included also another group of plants, the so-called "ruderal" plants. "Ruderal" plants recognized by the flora in those places which are in the vicinity of the dwellings and the settlements, by roads and by places connected with human activities and cattle breeding. The name of the group comes from the Latin word 'rudus' which means 'remains after fire, ruins'.

The seeds of these plants are very often supplied with different appliances for hooking with hooks and prickles which get into human clothes or in the fur and skin of the animals and thus get moved sometimes over large distances. The archaeophytes demonstrates high adaptability to environmental conditions and to certain cultural plants. They are characterized with:

— Easy dissemination - big part of the archaeophytes are adapted to easy dissemination in big distances through: wind, water, often through clinging to animals, fur and skin and by man when they stick to his clothes, etc.
— They have high abilities for spread – each plant formats a big number of seeds. Often they preserve their abilities for germination for quite a long time (4-7 years).
— Adaptability to unfavorable conditions – they develop successfully both by high humidity and by dry conditions, in rich and in poor soils and they are flexible to the soil reactions.
— Most often their seeds ripe before those of the cultural plants, they produce several generations in an year, they form seeds similar in size and form to those of the cultural species with which they adapted to grow (and that put obstacles in their clearing), etc. Different groups of archaeophytes are formed depending on their biological peculiarities: annual and perennial plants, early or late spring cultures, winter-spring ones.

By studying the charred plant remains among the cultural plants are very often found seeds of archaeophytes as well as series of wild growing species from which valuable information could be achieved for the economy of the households and the paleoecological conditions in the studied periods.

It is known that the weeds and wild growing flora is a good indicator about the environment. Both the archaeophytes and the wild growing species could provide an opportunity to establish what species were intentionally sown. As we all know, the winter cultures are wheat and barley. By the rotation of crops the wheat plant is sown in the autumn, most often after the fallow land, irrespective if it is green or black. The autumn cereals cultures are sown mostly following row crops or leguminous. By the black fallow land (the fallow land is black when the farmers does deep

plough in the autumn and from this moment on till the sowing in the field in the next year he keeps the land clean of archaeophytes which are not given the opportunity to germinate examples) develop such annual as the different species of: *Veronica cymbalaria*; *Veronica trifolia*; *Veronica dididma*; *Veronica arvensis*; *Veronica pelita*; *Stellaria media*; *Poa annua*; *Amaranthus retroflexus*; *Polygonum convolvulus*; *Chenopodium album* etc. If in the green fallow land there are cattle to graze ("green" is the fallow land when the land gets ploughed before sowing – very often there is green fallow land where the inhabitants does not have special lands for grazing so it brings the cattle to graze in the arable lands), there appear also such plants which have their adoptions to catch to and be moved by man and cattle. Such examples are the different species: *Xantium spinosum*; *Cynoglosum officinale*; *Cirsium lanceolatum* etc.

It is ascertained that some of them have vertical distribution in the sowings. They are of substantial significance concerning the yield. Often they enter in the archaeobotanical samples and thus they could provide valuable information about the way of harvest – if it was a high or low one. These archaeophytes are divided into three groups:

I level – here are referred those archaeophytes which grow higher than the cultural plants. This weeds is the most dangerous as divided by quantity they destroy the sowings. Example are: *Cirsium arvense*; *Melilotus alba*; *Lactuca scariola*; *Phragmites communis*; *Chondrila juncea*.

II level – it is much more widely spread. Could be found in almost every field and typical by for it is that it falls by harvesting in the sheaves, i.e. it gets cut by harvest together with the crops.

III level – to this group refer the plants which rest after harvest in the fields and then some of them grow up. The archaeophytes from the II level could be of interest to us as admixtures to the seeds of the cultural plants. Examples here included: *Agrosthema githago*; *Centaurea cyanus*; *Convonvulus arvensis*; *Galium apparine*; *Lolium temulenthus*; *Avena fatua*; *Chenopodium album*; *Vicia cracca*; *Vicia sativa*; *Vicia hirsita*; *Ranunculus arvensis*; *Polygonum aviculare*. The third level is the safest as the seeds of the archaeophytes here do not mix in the storaged yield as they grow lower and so they refer to the category of the harmless archaeophytes. Here we include again the different species of *Veronica* - speedwell: *Veronica cymbalaria*; *Veronica trifolia*; *Veronica dididma*;*Veronica arvensis*; *Veronica pelita*; *Anagalis arvensis*; *Ranunculus repens*; *Plantago lanceolata*; *Leontodon automnalis*; *Cerastium triviale* etc. They stay in the fields and by favorable conditions such as by fallow land or by row crops they grow in big number.

After the first year of use when the field ceases to be

ploughed and sown, is covered with typical archaeophytes whose behavior is different. After the second, third and so on following years some alterations in the archaeophyte communities appear still the vegetation does not return to its initial stage. It is known that the wheat-archaeophyte communities in the over time change. Impact on the process has also the harvest technique in the field, the type of fertilizing the soil, the grazing. For example if there was a low harvest the seeds of the low growing archaeophytes are also going to enter into the archaeological context, but if the harvest is too high many of them are not going to enter there. Thus also the archaeophytes adapt for dissemination among the wheat and they usually present a middle layer. This middle layer is much more widely spread. It is found more or less in every field and typical for it is that by harvest it enters into the sheaves, i.e. it gets cut by harvesting together with the intentionally grown cultural plant. On the other side the seeds of the plants which are higher or lower than those of the wheat spread spontaneously and they are in lesser quantities. The different archaeophyte communities by one and the same type of soil depend on the different agricultural technique.

The results of the studies of archaeophyte and wild growing grass vegetation provided rich information. As result of it 92 species of archaeophytes and grass weed were found. The most commonly found are: *Chenopodium album*; *Polygonum aviculare*; *Polygonum lapathifolium*; *Polygonum persicaria*; *Rumex acetosa*; *Rumex acetosella*; *Bromus secalinus*; *Agrosthema gitthago*; *Setaria viridis*; *Setaria glauca*; *Vicia tetrasperma*; *Vicia sativa*; *Vicia sp*; *Centaurea sp*. From these species highest is the presence of the following plants: *Chenopodium album*; *Polygonum aviculare*; *Rumex acetosa*; *Rumex acetosella*; *Agrostemma githago*. These species have wide ecological diapason and they could be met in different natural phytocenoses. Some of them such as *Chenopodium album*, *Polygonum aviculare* were found in the territories of the Balkans and in particular in the Bulgarian lands from the beginning of the Neolithic.

As in the archaeological contexts they were discovered quite often, some of them are presented here bellow together with their characteristic features as well as with their eventual use by man.

Agrostemma githago L. – corn cockle

Its seeds are triangularly rounded. Its size is: 2, 3 - 4, 0 x 2, 3-3, 6 x 1, 9 – 2, 9 mm. *Agrostemma githago* L. is an often found weed in wheat fields. The seeds get ripe in August. It is spread in the fields. The very young leave rosettes are used in some regions for filling in salted pastries. The seeds of the plant are poisonous for the birds. They contain up to 5-6% saponyn.

Bromus arvensis L. – rye brom

The plant grows up to a height of 90 sm. It is a weed of meadows, old vines, grow on poor soils, winter and spring wheat sowings. Its primary habitat is grassy places and secondary– sowings. The rye brom is an annual plant. Its seeds appear in the autumn and they pass the winter. It blossoms and gives fruit in wheat sowings from May to June, and among the row crops - from July to September.

Its origin is Oriental-Mediterranean. *Bromus arvensis* L is one of the most widely spread species of rye brom in the Bulgarian lands. It develops in sowings up to 900 m above the sea level.

Chenopodium album L. – fat hen

Its seeds are small rounded in outline with a few projections, with short radicile tips and clear, sharp border. The upper surface is with rounded; often the surfaces of seeds have specific reticulations that facilitate determination. The charred and non charred seeds of fat hen are very similar. Seed size varies from very small, les than 1mm to over 2mm.The plant is widely spread weed in the whole country. Its young leaves could be used for nutrition instead of spinach. Plant could be used for the production of red paint. Its seeds could be mixed with the fodder.

Cirsium arvense (L.)Scoop. – creeping thistle

Its fruit is cylindrical, laterally flattened, tapering below, faintly angular and slightly arched. It has marginal achenes which are greatly bowed. Its upper end is oval, usually slightly oblique, with smooth neck and central style residue. The surface of *Cirsium arvense* (L.) Scop. is smooth or with very longitudinal stripes, mat and brownish as color.

Its seeds grow ripe late in summer. Size: 2, 0 – 3, 0 x 0, 9 – 1, 2 x 0, 7 – 0, 8 mm. It is found in the fields and in the rocky, ruderal and humid places.

Fallopia convonvulus (L.) A. Love*, syn. Polygonum convolvulus* –black bindweed

Fallopia convonvulsus (L.) has a fruit with three sides, pointed at both ends. Its broadest part is in the middle. The edges are slightly rounded. Lateral surfaces indented, especially in fruit that is not completely ripe. Its surface is similar to that of Knotgrass (*Polygonum aviculare*) - with small bead like nodules in longitudinal rows. Its seeds are black to brownish-black. Their size is: 2, 7 – 4, 3 x 1, 8 – 2, 7 x 1, 8 – 2, 7 mm.

The habitat of the black bindweed is the bushes and the grassy lands. Its fruit contains proteins. It is often find in spring sown, meadows. Its leaves are nutritious; the plant prefers humus, moist. Collected for food are only the young leaves. Fresh the plant is suitable for cattle grazing.

The seeds are good fodder for the poultry but they also get used for nutrition prepared as grits (manna croup) or added to wheat flour. Their seeds turn the bread into dark colored and it achieves a specific taste. They get also mixed with straw to be used as animal fodder.

Polygonum aviculare L. - Knotgrass

The length of its seeds varies from 1, 0 to 3, 00 mm. After the harvest its seeds hibernate. The knotgrass can been found in spring-sown crops. It prefers carbonate soil.

Saponaria officinalis L. – soap wort

It has rounded kidney-shaped seed rather flatted around the hillum with rounded edges. The surface of *Saponaria officinalis* L. has concentric rows of blunt, smoth, elongated protuberances. Its hillum is deeply sunk, surrounded by a projecting, rather tuberous margin.

The seeds of soap wort - *Saponaria officinalis* L rippen in late summer. Their size is: 1, 5 x 2, 2 – 1, 3 – 1, 8 x 0, 7 – 0, 8 mm. *Saponaria officinalis* L. grows in humid and stony places. The whole plant but most its roots contain saponin. The dry substance of its roots there is about 13-15% of the poisonous saporubin. The plant is poisonous for the cattle. Still it is used in traditional medicine. It is also used for washing.

Centaurea sp. – star thistle

The size of its fruit varies significantly- the smallest hardly reaches a length of 0, 5 mm while the biggest ones often could be 9 mm long. This family is spread all around Bulgarian lands as weed plant. *Centaurea sp.* is a typical weed in the winter wheat sowings. It develops well in slightly-crumbly and rich in nutritious rich substances soils.

Vicia sp. – vetch

The seeds are rounded or oval, sometimes flattened/squashed, the seed edge could be rounded, oval, elliptical, elongated or short. By determination of seeds of different *Vicia* species the seed edge appears as important systematic feature.

Lythospermum arvense L. -field gromwill

The fruit is slightly triangular, irregularly egg-shaped. The seed length is 2, 0 – 3, 0 mm and its width is 1, 5 – 2, 2 mm. The plant is found as weed in the fields in different weedy lands, mostly in the winter wheat and rye sowings, in the spring sowings by the oats, millet, lentils and peas.

Rumex acetosella L.– cheap's sorrel

Its seeds are small, triangular and they are slightly sharp-ened in their both ends. The sides are equal in lenght. It is a wide spread weed.

Conclusions from the Study of the Weed Flora

The study of the wild growing vegetation provides positive results thanks to the implementation of the flotation method. The results of the study are show in fig. 35. It is an attempt to classify the sowings according to the character of the accompanying weeds. This is not always very precise as there are typical weeds for spring and for winter sowing, but some of them keep- intermediate position. The early spring weeds grow in big quantities in the spring and they grow mostly among the cereals and the early spring cultures. The winter-spring cultures enter as weeds, the winter wheat and the early spring cultures. The winter weeds start its development in the autumn and it forms seeds before the harvest. They require long hibernisation. They grow as weeds in meadows, grazing lands, perennial grasses and wheat-corn cultures. In that connection from fig. 35. is seen that in the bigger part of the studied settlements are established winter or winter-spring sowings but, having in mind that they weeds the winter wheat cultures it could be eventually presumed the existence of winter corn sowings. But on the other part, some of the weeds are typical also for the early spring sown so it is difficult to do a more precise evaluation of the situation. It is possible also that there existed winter wheat sowings and spring sowings around the location with access to water for vegetables and pulses. And on the other side the found weeds could help to ascertain defined agricultural regimes in the past.

From phyto-sociological point of view a part of the weeds vegetation is divided into two main groups: *Secalinetea* and *Chenopodiatea*. The first group is typical for the mostly winter (wheat) sowings and the second one - for the gardens and pulses Gones (1992, pp. 133-143). Participants of both groups are found in the studied settlements. Thus in the settlement of Yabalkovo - 3 species weeds are typical for the row-crops and they belong to *Chenopodiatea* - *Bromus secalinus*; *Chenopodium album*; *Rumex sp.*, but there is also a representative of the second group *Secalinetea* - *Agrostemma githago*. In the settlement of Madrets the species of the both groups are equally represented while in Slatino the weeds representation is inclined more to *Chenopodiatea*. The quantity of the weeds species characterizing both groups is not quite sufficient for going out with definite assumptions. If an analogy with some studied settlements in Greece is done, then according to Perles (2001, pp.164-165) the cereals was sown in autumn but there was also the practice alongside the rivers of spring sowing as well. For sure a bigger representative sample of the wild growing flora is needed to provide us with more entire information so that we could define the regime of cultivation.

4. FRUIT AND WILD GROWING PLANTS – GATHERING

Together with the remains of cultural plants the archaeological sites often are abundant with remains of wild growing plants and fruits. The information about the gathering in the past of wild growing plants is only partial and mostly based on the findings of their seeds and stones. In most of the early settlements the wild growing plants were seen as providing a significant parallel source of food. The findings of fruit bearing trees remains could be divided into two groups: those found in the territory of Bulgaria so they are of local origin and another group of species which are typical for the Mediterranean regions.

The following species refer to the first group:

Quercus sp. - oak

One of the most widely spread tree in the territory of the Bulgarian lands – the oak in its different varieties – is known here by the remote past. The numerous oak species constitute the main plant formation in the Old World – as in the regions with more moderate climate so also in the Mediterranean regions. The ever green species *Quercus coccifera* L. and *Quercus ilex* are typical for the Mediterranean vegetation. The *Quercus robur* and *Quercus petraea*(Matt.) Liebl. are widely spread in the European regions with moderate climate while *Quercus brantii* Lindley and *Quercus ithaburensis* Decke are dominants of the forests in the Near East (Zohary and Hopf, 2000). Most of the oaks at the end of the summer produce big quantity of acorns. The size of the acorns is dependent to a big extent on the temperature conditions. In the moderate zones the acorns are smaller in size compared to those in the Near East where the acorns reach to a length of 5-7 sm with a width of 2,0 – 2,5 sm.

Several oaks could be found in the territory of Bulgaria. Each of them requires different climatic conditions.

— *Querucs coccifera* species is found on dry hills in the warmer parts of the country – in the vicinity of the towns of Petrich, Sandanski, Dupnitsa. In Sothern Macedonia and in Western Thrace this species is the main source of hard fuel.
— *Querucs cerris* L. species grows mainly on the hillsides by the high mountains in the country.
— *Quercus conferta* Kit. could be found not only in the lower parts of the mountains but also in the lowlands near the Balkans.
— *Quercus sessiliflora* Salisb. – "winter oak" – could be found both in the lowlands and in the lower parts of the mountain ranges.
— *Quercus pedunculata* Ehch. – "summer oak" species grows in the low lands and alongside the rivers.

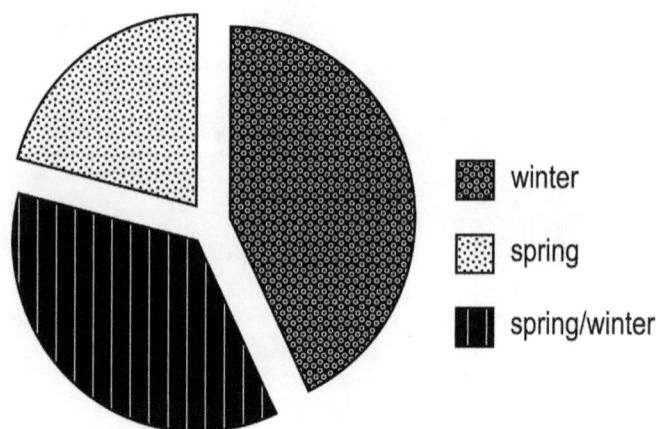

Fig. 35. The percentage correlation of the different sowings in the studied settlements.

The acorns were frequently used as alimental addition mostly in the autumn. When there was shortage of wheat or loss of wheat yield acorns were added to the wheat and milled together thus turning into flour and further into bread. All acorns contain tannin but in some the concentration of tannin is lower as is the case with those that originate from the Mediterranean regions. The acorns are often used as preferred food in swine breeding.

Charred remains of acorns are often met in the context of the Neolithic and the Bronze Age in the Near East, in the Mediterranean basin and in the Balkan Peninsula.

Acorns findings presented in many settlements as separate fragments or halves of fruits or even as whole fruit. Of special interest are the storages of about 4 kg acorns found by the tell settlement Hotnitza in the region of the city of Veliko Tarnovo Popova (2008, pp.189-194). Most probably they were meant for swine nutrition as in the same settlement sufficient storages of wheat also was found. Certain quantities – probably as alimental supply - discovered in the tell settlements by the village of Yunatsite and by the village of Dyadovo – dated from the Bronze Age Popova (1991b, pp. 69-72). Hajnalova (1975, pp. 303-314)) announces a big quantity of acorns found in the tell settlement of Golyamo Deltchevo.

Prunus avium (L.) Moench – Cherry tree

The cherry tree's habitat is often in the vicinity of the settlements but it could be found everywhere in the country. According to Franke and Hammer (1976: 108) the cultural cherry probably originates from regions connected with the Black Sea. It is known back about 8000 B.C. in Anatolia as well as in some Neolithic pile dwelling settlements in Switzerland. It is known that in 74 B.C. L. Lucullus (about 73 B.C.) was the first to introduce it to Italy.

The wild growing cherry tree "*avium*" could be connected with some fruit stones found in Neolithic settlements layers as well as in such of the Bronze Age - Durankulak - Popova and Bojilova (1998, pp. 391-399); Popova (1995a, pp. 261-266); in the tell settlement Dyadovo Popova (1992, pp. 238-246) and in Adata, Koprivlen - Bronze Age (Popova, 2003).

Pyrus communis (syn.P. domestica Med.) - Pear

The cultivated pear is the second fruit bearing tree introduced in Europe. The family has about 20 species spread in Europe and in Asia. There are no precise data on it exact origination. Some of its varieties are spread in the Ponthic region and in Kolhida but in China wild growing pear could also be found. Some of those species should have produced through hybridization the cultural pear tree.

The pear fruit has been object of gathering as wild growing fruit long before being cultivated. Findings of it were discovered in Neolithic layers and in the Bronze Age in Europe – in Swiss, in former Yugoslavia and in Greece. The archaeological evidences could still not provide sufficient evidence when it was cultivated exactly. To some extent partialy reliable information is achieved by the ancient Greek and Roman authors, mostly from Theophrastus, who described three cultural varieties for Greece while Cato described its method of cultivation. In Bulgaria findings of pear were found in the settlement Dana Bunar, Dimitrovgrad district - Chalcolithic period (unpublished data), in the tell settlement Dyadovo - Bronze Age, (Popova 1992, pp. 238-246) and later in the Roman epoch in the roman settlement Drenkovo, Blagoevgrad district (unpublished data).

Prunus domestica – Plum tree

It is found in the forest where it grows by the outskirts of the forests. The cultivated plum tree – *Prunus domestica* – is a polyploidy form, result of the crossing of *P. cerasifera* with some species of *Prunus*, which lately spread in Central and Northern Europe.

Pits of *Prunus domestica* were recovered in sites of the Neolithic and Bronze Age in the territory of Swiss, Italy and Germany. In the territory of Bulgaria *Prunus domestica* were also recovered in the Neolithic settlements of Durankulak, Dabene, Kapitan Dimitrievo, Karanovo, Kovachevo, Koprivets, Samovodene, Bronze Age settlements - Galabovo, Madrets and of *Prunus spinosa* L. – in the Neolithic layer of the settlement Durankulak.

Cornus mas L. - Cornel-tree

The cornel-tree species - *Cornus mas* – is either a bushes or a small tree. Its habitat is Southern and Central Europe, the Black Sea region and the Caucasus Mountains.

The cornel-tree is found in the bushes, in the forests and

in the rocky places, up to a height of 1300 m above the sea level. Its wooden fibers are of exceptional strength.

The fresh cornel-tree fruit has high contents of Vitamin C. Its bark, branches and leaves contain yellowish pigments and for that reason they are used for coloring.

Cornel-tree fruit stones are found in numerous Neolithic and Chalcolithic settlements in Europe and in particular in the Balkan Peninsula. In the territory of former Yugoslavia the earliest findings dated are of the settlement of Opovo (Borojevic, 2006); and in Greece – Macri – from the middle and late Neolithic (Valamoţi, 2004).

Evidences of cornel-tree mark the entire pre-historic epoch in the territory of the country. They are documented in the sites: Kovatchevo – early Neolithic; Kapitan Dimitrievo – early Neolithic; in Durankulak, Galabnik, Karanovo III-IV, Koprivets, Ovcharovo, Samovodene, etc. In many of those sites charred cornel-tree fragments and sometimes even whole fruit stones of *Cornus mas* were found.

Sambucua nigra L. - Elder

The Elder is found in the bushes and in the forests and quite often its habitat is around the settlements.

Its fruit (*Sambucus nigra* L.) is small sized. Usually it contains 3 seeds. The size of its seeds varies from 3,0-5,0 mm length to 0,7-1,2 mm thickness (Schoch, et all., 1988: 35). Its blossoms (of *Sambucus nigra*) contain about 0, 025 % ether oils, among them turpentine and paraffin oils. The elder is one of the most popular plants used in the traditional medicine. All of its parts come in use. Its blossoms is known as a very good remedy that helps in cases of 'flue and of bronchitis. The *Sambucua nigra* L. fruit is used for the preparation of syrup as the fruit is abundant in the vitamins of A, C, C2. The fruit also gets used for preparation of blue color for painting cloth-tissues.

In Europe findings of elder remains were discovered in the Neolithic settlements. Such findings are not rare in the territory of the Balkan Peninsula. In the Neolithic the findings in most cases are single ones.

Elder was found in the Neolithic layers near to the currently village of Vesselinovo (Arnaudov, 1947-1948). In the Chalcolithic layers of the tell settlement Galabovo they are also present. In the tell settlement Yunatsite a significant quantity of elder (*S. nigra*) was found and without any admixtures Popova (1991b, pp. 69-72). It is fixed as finding also in all studied periods of the tell settlement in Durankulak – in the Neolithic, Chalcolithic and the Bronze Age Popova (1995b, pp. 193-207) (fig. 36).

Sambucus ebulus L. - common elder

Common elder is found to grow in the warmer parts of the regions with moderate climate in Europe. It is a smaller grassy plant which grows in fertile humus lands; it is found mostly by roads cut in the forests. It could be found by waste lands and alongside rivers as ruderal plant. The common elder is a small succulent fruit usually with two or three seeds. The seed surface has furrows. The sizes of the seeds vary in the limits of 0, 2 - 0, 35 mm length, 1, 5-2, 3 mm width and 1, 0-1, 5 mm thickness (Schoch, et all., 1988:35).

The common elder indicate of fertile soils. Plinius mentioned it as testimony for best soil conditions and that as weed it was known to the Romans. He recommends it to be used for coverage of the floor by breeding cattle. The common elder was used in the traditional medicine from the ancient times. All parts of the plant could be used against bronchitis, kidney troubles, and snake-spider-bee bites. The fruit is used also to provide nice red color to wines and as supplement in marmalades.

In Europe its remains were found in Neolithic settlements. The Balkan Peninsula such remains are recovered in Kastanas – Greece (Kroll, 1979), in the Neolithic settlement Selevač and in the Neolithic settlement Opovo in former Yugoslavia (Borojevič, 2006). The plant exists in the archaeological sites back from the Neolithic – it was found in the Neolithic layers in the tell settlement near Vesselinovo, Kapitan Dimitrievo, Karanovo II-III, Karanovo III, Koprivetz.

Vitis vinifera L. – (wild) vine

The plant inhabits in the mesopfile forests and bushes by swamps and rivers. In our lands it grows alongside the Danube river and the Black Sea coast, in South-Western Bulgaria, in the Thracian lowlands, in the valley of Struma and in the Eastern and Middle Rhodopes. The wild vine is resistant to drought and cold.

Vitis sylvestris Gmell. is widely spread in Europe and in Western Asia. Its pips differ in form and size, grape of the wild vine (*Vitis vinifera ssp. sylvestris* Gmell.) is rounded, thick, with shorter stem/stalk of the grain. The cultivated grape (*Vitis vinifera ssp. vinifera*) is elongated, fragile and flattened. The size of the pips of *Vitis vinifera ssp. vinifera* are between 3, 5 – 6, 00 mm long, 3, 0 – 4,1 wide and 2, 2 – 3, 0 mm thick (Schoch, et al., 1988: 96).

The fruit contains 22-23% sugar.

Charred pips of grape are found in numerous pre-historic archaeological sites in Europe. The fruit of the wild vine were collected long before the plant was cultivated. The findings in Greece, former Yugoslavia, Bulgaria, Italy and Swiss are from Neolithic sites and these early materials judged on the basis of the morphological features of its

pips correspond to these of the local wild "*sylvestris*" species of grape (Renfrew, 1973).

In the region of the Aegean Sea seeds of the grape which come close to the cultivated species "*vinifera*" according to the index width to length are found in the early and middle layers Kelladic of Lerna (Hopf, 1961). Grape remains are reported also from the early Minoan Mirthos in the island of Crete (Renfrew, 1973) as well as of Troya and Beyçesultan in Western Turkey (Helbaeck, 1960). In Sitagroi, Northern Greece, Renfrew (1979, pp. 243-265) studied series of pips dated from between 4500 and 2000 B.C. She ascertained a definite transition from the "*sylvestris*" species with short sharpened pips in the early levels towards a species similar to the "*vinifera*" in the late layers. Thus she presumed that the vine growing in Macedonia had appeared quite earlier than 2000 B.C.

In the Balkan Peninsula *Vitis vinifera ssp. sylvetris* is found in layers of the late Paleolithic and Mesolithic in the Franchthi cave in Greece. Kroll (1981, pp. 161-171) also found wild growing grapes - *Vitis vinifera ssp. sylvestris* in series of sites in Greece, dated from Early to late Neolithic. Pips of wild growing grapes are found also in Neolithic sites in the territory of former Yugoslavia as well as in Bulgaria.

The earliest findings in the territory of Bulgaria refer to the Neolithic sites in: the village of Malak Preslavets "Pompena Stanziya', the village of Drinovo, in the villages of Eleshnitsa, Kapitan Dimitrievo, Karanovo (cultures Karanovo II – III), Orlovets – Neolithic. All of them refer by morphologic characteristics to the wild growing vine. Finding of cultural vine appear in the Bronze Age in the tell settlement near the nowadays village of Yunatsite and in the archaeological sites Yazdach, Koprivlen and Nebet Tepe (fig. 37).

Rubus L. - blackberry, raspberry

The family Blackberry has more than 300 subspecies and many hybrid forms. Most often found species are: *Rubus fruticosus*; *Rubus idaeus*. The blackberry bush is widely spread in the moderate climate zones in Europe, Asia and America. Its fruit is small and succulent, consisting of multiple fruit fragments with a single seed. It grows in humid and rocky places. The fruit gets ripe in late autumn. The leaf of the Rubus L. contain tannic substances used in the past in the traditional medicine. The young leaves are used for tea.

The archaeological findings of *Rubus* L. species remains are dated in the Neolithic – Opovo, former Yugoslavia (Borojevic, 2006), Kastanas – Greece, and from the early Bronze Age (Kroll, 1979). The data from the studied plant remains show findings of *Rubus* L. species remains in the tell settlement of Chavdar near to the town of Kazanlak, in the tell settlement of Karanovo (Karanovo II – III), etc.

FRUITS WITH MEDITERRANEAN ORIGIN

As result of series of archaeological and archaeobotanical studies enough quantity of material which marks certain species met in the territory of the country as well as some imported species has been found. These plant remains were found only in the last years and then they were witnessed for first time in the territory of Bulgaria. The finding of remains of *Pinus pinea, Pistacia terebinthus, Ficus carrica* L. are findings of remains of plants with Mediterranean origin.

Ficus carrica L. - Fig-tree

The fig was imported from Palestine and Mesopotamia to Egypt and Greece. The earliest fig remains are found in the settlement Jericho (7000 B.C. – aceramic Neolithic, layer A and in 5000 B.C. – Chalcolithic. In Europe the fig is documented in the cave Grotta del'Uzzo – Mesolithic – Neolithic (Constantini, 1981). In our lands for first time the fig appears in the Bronze Age in the settlement Galabovo Popova (1995a, pp.261-266) and in the settlement Madrets, Popova (1994, pp.111-119). New data show the earliest finding of the Neolithic layers in the settlement Jabalkovo, dating of Neolithic period.

Till recently some authors thought that this species was introduced in our lands during the times of the Greek colonization. Our latest data proves that the fig either existed permanently in the warmer regions or it was introduced much earlier.

Pinus pinea L. – stone-pine/Italian pine

The *Pinus pinea* L. grows by warm and dry conditions. No the north it harder gets acclimatized and develops fruit for what it is logical that it has moved to our lands from the Mediterranean region.

The stone-pine (*Pinus pinea* L.) is a typical Mediterranean plant which could be found in the vast area from Syria to Portugal. The collection of cones what its fruit is seems to have been quite usual practice in the Near East as well as in Cyprus and in Greece. The tree was widely spread even before the time of Ancient Greece especially in Italy where it had been used not only for culinary and decoration but also even for rituals. Many authors - Plino, Columella, and Paladius mentioned its consumption. Most often is could be met in the receipt of Apicius (I century B.C.). Its findings mark the entire Roman epoch in Western Europe.

The remains of a sunken ancient ship discovered in Ulu Butum (near to the Turkish Mediterranean coast) were dated from the Bronze Age (about 3200 B.C.). The boat trans-

ported huge quantity of cones of *Pinia pinea* L. which provides evidence of the importance of this fruit in the trade of the time Marinval (2001, pp. 108-109). The findings in Pompeii and Herculean witness the common consumption of this species. Its strobiles were burnt in ritual actions connected with the fertility. The repetition of its findings shows its significance as a ritual connected with the future life of those who passed away. Judging by the context of the remains that have been found it is obvious that also in the territory of the Bulgarian lands *Pinia pinea* L was used both for consumption and in rituals.

Its earliest finding in the territories of Bulgaria is dated from the Bronze Age. The discovered seeds of pinia (*Pinus pinea* L.) are from a ritual hearth in the Thracian sanctuary complex Tatul. But because of the singularity of this finding in the recent paper any suggestions and guesses are to be omitted.

Later the *Pinus pinea* L. was established as finding also in Abritus (I-IV century), in Kabile and Sozopol as well as in the Gyaurskata Mogila near to the town of Karnobat – early Roman time.

CONCLUSIONS OF THE WILD GROWING PLANTS

From a review of the described findings an interesting fact was ascertained. The data of the collection show the usage of three basic wild species: cornel-tree fruit, grapes and elder. In Fig No 26 is presented the spread of the cornel tree in the different periods according to what the presence of the cornel tree in the entire pre-historic epoch could be traced – the highest percentage in the settlements evidenced in the Neolithic. The findings of cornel tree in the Neolithic show high presence of the species of *Sambucus ebulus* while in the Bronze Age the *Sambucus nigra* species is prevalent. The third wide spread plant species was the vine. Its presence in the different archaeological sites is shown in fig. 27.

Differences in the spread in time of the cultural and the wild growing species are observed quite clearly. The wild growing species are predominant in all Neolithic and Chalcolithic, while in the Bronze Age already noted the typical features of its domestication.

5. PLANTS FOR TECHNICAL USE, MEDICAL PLANTS AND PLANTS WITH TOXIC SUBSTANCES

Except for nutrition very often the wild growing plants were used also for other purposes. On this basis they could be divided into several groups:

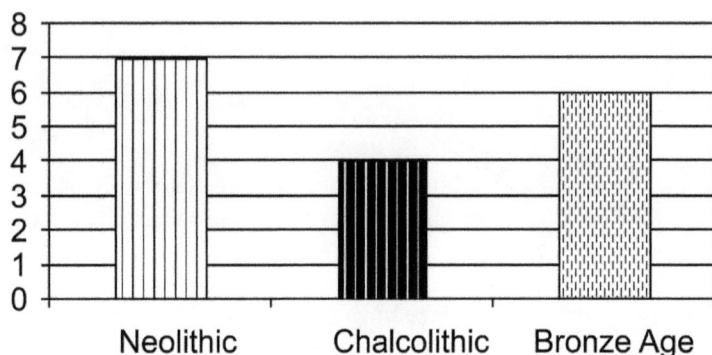

Fig. 36. Percentage correlation of the cornel tree findings in the studied settlements by periods.

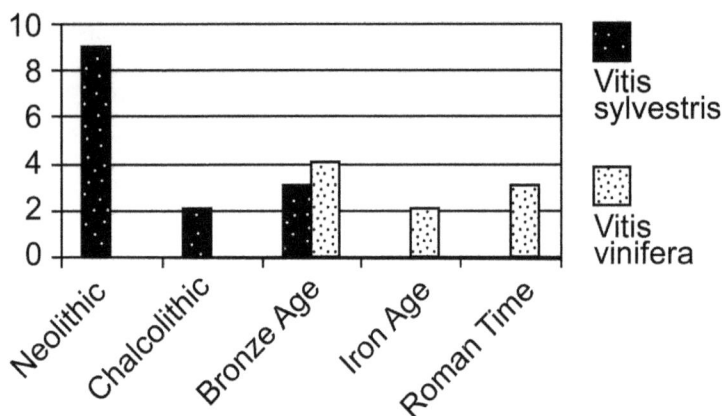

Vitis sylvestris

Vitis vinifera

Fig. 37. Percentage correlation of the will growing and cultural grape in the studied settlements by periods.

— Animal food / fodder /forage
— Use in construction and carpentry
— For technical use

The origin of the anhtropophyte vegetation is often explained as developed around the man dwellings and around their settlements. The environment which man creates with his activity impacts mostly the soils features. By agricultural activities the natural vegetation gets damaged and replaced by anthropophyte vegetation. The anthropophytical plants have developed after suitable conditions for them were already created, and namely after the different kinds of treatment of soil and the change in the natural environment that followed. Much later man started to cultivate plants for technical and other purposes. Part of the wild growing flora had a definite role as supplementing resources and it was used fully adequate. Different parts of many plants were used as aliments by man himself or by his animals. Presented in the figs. 39-40 bellow are some of the most often found species which came in use and some of them still stayed in use.

In the first group are included the seeds of some species such as: *Amarantus sp.* different species of *Setaria* - *Setaria italica*; *Setaria glauca*; *Setaria viridis* and different species of Vicia - *Vicia sativa*; *V. angustifolia* – which were used as food for the poultry. Except for those listed them some species - clover, vetch - *Trifolium sp*; *Vicia sp.* appeared as good fodder for the animals.

In winters in regions with forests leave fodder was also used for the nutrition of the sheep and goats. The oak leaves were preferred as well as those of hornbeam, beech, etc. which got cut in the early autumn.

The second group comprises plant species used in construction. Most of the wood used in construction was of the easily accessible species such as those found at the ends of the forest. Thus the hazel and the cornel trees, some lianas as the clematis whose wood is elastic were used in the weaving of fences or for the basis of walls that were later clayed. The more robust wooden material of ash tree or of different species of Rosaceae was prepared objects used in everyday life. The wood fiber of lime-tree, hazel-tree and clematis was used also for weaving baskets. The straw and the chaff were added in the clay by the erection of different constructions for it were those of the emmer was very suitable robust and less breakable.

The third group – Though not very common, in the archaeobotanical materials from the studied settlements some plants were established which, eventually could have a definite use for technical purposes. Part of them is presented here.

Linum isitatisum L. - linen. The findings of linen in the archaeobotanical materials are rare, and some of them include the wild growing species. The earliest for now refer to the settlement of Slatina Docheva (1992, pp. 144-152). So far the plant does not have wide dissemination in the

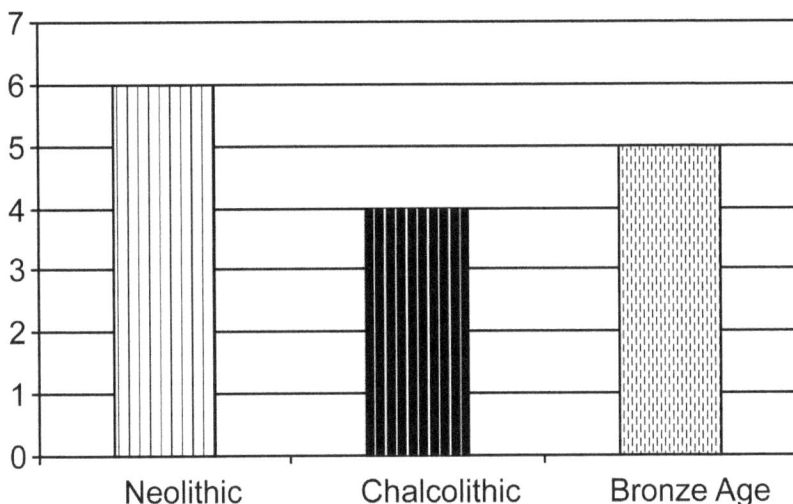

Fig. 38. Percentage correlation of the fat hen findings in the studied settlements by periods.

archaeobotanical materials. Single findings are witnessed in Kapitan Dimitrievo, Slatina Marinova (2002a, pp. 1-11) and in Karanovo Thanheiser (1977, pp. 429-477). The linen seeds have good conservation and probably it was the cultural species of the linen.

Chenopodium album L. - fat hen, *Sambucus ebulus* L. - common elder, *Sambucus nigra* L.- elder. These plants were used for coloring in dark blue. Additionally the plants were used against bronchial and cold troubles. There are data of conservation. The fat hen is one of the most widely spread plants and some authors are on the opinion that it was intentionally collected. It is evident that it is often presented in the studied settlements. The frequency of its presence is shown in fig. 38.

Of interest are the found imprints of bracken fern in the daub of a pithos in the Hotnitza. The bracken fern is found in the bushes and in the forests. It is known that the leaves were used for level covering in the cattle-sheds, where they improve the quality of the manure. Thanks to the specific smell of the leaves and also due to their anti-decay qualities these leaves are still and used in the past for wrapping the fruits and vegetables. It is quite possible that the bracken fern was used also for the conservation of some nutritious products.

Urtica dioica L.- nettle - the name of the plant originates from the Greek word "kopros". This plant is an old cultural thermophyt. Grown for nutrition but later also as oleaginous and medical plant. It is very common and indicates nitrogen rich ground. Widely spread everywhere in the country mostly by waste places and fences. The nettle characterized very well local environment.

Cannabis sativa l. - hemp, cannabis

Till recently the *Cannabis sativa* is rarely presented in the archaeobotanical findings. It was found in the form of texture and, in most cases, in later periods. *Cannabis sativa*

in the form of wick is found in the settlement Ezero, dated from the Bronze Age Arnaudov (1949, pp. 87-116). Its sole finding in a pre-historic site is dicovered in the settlement of Topolnitsa (unpublished data).

OLEAGINOUS AND MEDICATIVE PLANTS

Brassica nigra – black mustard

The *Brassica nigra* L. is presented in two archaeological sites – in Durankulak and in Kamenska Čhuka Popova (1999a, pp. 478-481). It is an annual grassy plant grown for its seeds which are used as a spice. It is accepted that the plant originates from the southern Mediterranean region in Europe. The spice is usually prepared by the milled ground seeds of the plant. The small (1 mm sized) seeds are hard and colored in dark brown to black. They have a strong taste though they almost do not have any aroma. They contain a significant quantity fat. The produced by it fluid is often used as cooking oil.

Lithospermum arvense L. – field gromwill

It is found in the grassy lands and as a ruderal plant. Its seeds contain oil and iodine. They are used as aliment in poultry. The plant is documented in the following archaeological sites: Durankulak, Dabene, Kapitan Dimitrievo, Karanovo and Yunatsite.

Other species of this family are found as well in Durankulak - *Lithospermum officinale* L. and *Lithospermum sp.* in Kovachevo. Most probably the plant was used as food for poultry but it has also usage in medical treatment.

Proof of its implementation as such is documented of a finding in Szarbia, in the district of Koniusza, Malopolska province, in Poland. The grave of a woman was discovered there –dated from the Bronze Age, Mierzanowice culture (1750 - 1600 B.C.).

OILCONTENT PLANTS	FIBROUS PLANTS	MEDICAL PLANTS	TOXIC PLANTS
Brassica rappa	Cannabis sativa	Artemisia absithium	Agrostemma githago
Cannabis sativa	Linum issitativum	Sinapis arvenis	Atroppa bella donna
Linum issitativum		Verbena officinalis	Artemisia absitnhium
Urtica dioica		Sapponaria. officinalis	Lolium temulenthum
			Ranunculus acer
			Senecio jacobea

Fig. 39. Different use of wild growing plants.

species	man	animals
Chenopodium album	leafs	
Convonvulus arvense	leafs	fodder
Echinochloa cruss - galli		seeds
Galium apparine		seeds
Lithospermum officinale		seeds
Polygonum aviculare	leafs	fodder
Rumex crispis	leafs	fodder, seeds
Rumex acetossa	leafs	fodder
Rumex acetosella	leafs	fodder
Rumex sp.	leafs	fodder
Sanguisorba minor	leafs	
Salvia officinalis	leafs	
Setaria viridis		seeds
Setaria glauca		seeds
Trifolium pratense		fodder
Veronica hederifolia		fodder
Vicia ervillia	seeds	fodder
Vicia sativa	seeds	fodder
Vicia angustifolia		fodder

Fig. 40. Use of different parts of the plants.

In the very grave are documented numerous fruit of *Lithospermum sp.* For which the authors Baczynska and Litynska-Zajac (2005) presume that it was used as antiseptic medication.

Sapponaria officinalis L. - soapwort

The earliest findings of this plant in the territory of Bulgaria are from the sites Yabalkovo and Durankulak. It is a perennial grassy plant with a thick reddish root. Soapwort contains three-turpentine. The plant is used by bronchitis and troubles of breathing system. *Sapponaria officinalis* is poisonous. Both its roots and its twigs come into use, collected in autumn after its seeds got ripe. The plant could be found in grassy places, in bushes, by roads and fences.

Sinapis arvenis L. - field mustard, the plants is finding only

in the Bronze Age settlement - Dabene Marinova (2003, pp. 499-504).

Verbena officinalis L.- common vervain, verbena – The plant was found in the Neolithic site Balgarchevo (Marinova, E., E. Tchakalova, et all. 2002). Its habitat is grassy and weedy lands by roads. *Verbena officinalis* is spread throughout the country at a height from 0 to 1000 m above sea level. Used are its twigs. It contains the glycosides: verbenalin, etheric oil, tannic substances, turpentine, and resin oil.

PLANTS WITH TOXIC SUBSTANCES

Many of the pulses such as bitter vetch, broad beans and grass pea contain poisonous substances. Therefore, before boiling these, they should be soaked for these substances to be extracted. Some of them cause illness and the illness is called 'Phavism'. The illness was known back in the Classic epoch, it spread to Mediterranean region proven to be a hemolytic anemia – deficit of glucose 6 phosphate – dehydrogenate (G 0 6 – PDH), the so-called enzyme in the red blood vessel. The "Phavism" is genetically inherited and the one who has a defined recessive gene is susceptible so he reacts to the broad bean. In such plants are contained also not-alimentary content that could also provoke symptoms of hallucinations. Plants which cause "Phavism" are the broad been, grass pea. i.e . That is the reason they should soak in water before consumption – their toxins to get out.

Agrostemma githago L. - corn-cockle. The plant appears quite frequent as weed in the wheat. It is found in some of the studied sites – Galabovo, Yunatsite, Yabalkovo and i.e. The seeds of this plant contain toxic glycosides which bring narcotic effects.

Atropa belladonna L. - Belladonna, Deadly Nightshade

This plant has been found in the Bulgarian lands solely by Thanheiser (1997, pp. 427-477) in Karanovo. The Belladonna (*Atropa belladona*), also called "crazy herb" or "old herb" is a perennial bush from Solanaceae. It is found in whole Europe. Its typical habitat is shadowy humid places with rich in limestone soil. The Belladonna has spherical green leaves and lilac blossoms. Its fruit is black and shiny with a diameter about 1 sm. It could reach a height of about 1 m. The belladonna is a very toxic plant containing the alkaloid atropine. Most often its root is the most poisonous part of the plant.

Artemisia absithium L. – an ancient medical plant. It is very popular in the traditional medicine. It could be found in weedy places in the vicinity of the settlements and by the dwellings.

Plants with poisonous action such as *Valeriana celtica ishaemum, Helleborus odorus, Convolvulus cantabricus* are mentioned by Herodotus for the Thracian lands, but as they were not found in the archaeological contexts we presume they should not be discussed here.

The usage of some of those wild growing plants is presented in figs. 38-39.

Chapter 8
Different Types of Wood

Wooden resources – interpretation of the material

- — Concentrated wood
- — Dispersed wood
- — Contextual analysis
- — Taphonomic analysis

To studying charred wood and interpret the findings some important factors should be taken into consideration.

- — Fragmentation of the material
- — Comparison between different stratigraphic structures
- — Size of fragments
- — Number of fragments
- — The different contexts – concentrated and dispersed fragments
- — Quantity of the fragments in the stratum
- — Duration of use of the different archaeological structures
- — The diversity of the species

For example:
"The pits and different structures connected with burning characterize the type of the archaeological evidences as well as their presentation – affluence of taxons or insignificant participation" Heinz (1990, pp.20-21).

"The diversity of taxons is a function of the way of use of a given structure and its duration – long or short-term" Figeral (1992, pp. 321-362). On the other part from the study in each archaeological site quite heterogenic special distribution of taxons in each level of habitation is observed (Figeral 1990, pp. 37-38). It is established but that the basic species are found in almost all studied contexts where dominant is homogenous special distribution, while the heterogenic special distribution of the other taxons is found around and in even wider boundaries Figeral (1992, again there).

The charred wood originating from pits, hearts, and ovens, as mentioned above is called **concentrated**. The wood originating from different layers horizons, floors is called

dispersed. These two categories provide opportunities for studies in two aspects: palaeo-ecological and palaeo-ethno-botanical. In that connection some examples will be given where these both categories are applicable.

Palaeo-botanical aspect

Part of the material studied in the archaeological sites could be referred to the so called *concentrated* wood, i.e. were found from different archaeological structures mostly pits, hearts and ovens. An example of wood studied in this way is a hearths from the site Madrets (Bronze Age). The samples from the hearths are taken from different consecutive depths by directly dismantling it.

The presence of 4 wood species: oak, ash-tree, maple and hornbeam was established. These woods are used for different household needs – cooking, warmth, i.e. they are connected with the everyday life.

A similar example is the analysis of the filling of Pit No 1 from the sanctuary Adata. The general feeling is of a feeble representation of the species diversity but the qualitative representation is high. The dominant species is the oak with single fragments of hornbeam. The presence of coniferous species is documented only by 1 pine fragment – of *Pinus cf. sylvestris*.

There are not established grains and seeds – obviously in the pit there had been depositing of material exposed to fire lately.

Palaeo-ecological aspect

An example of **dispersed** wood is the study of the stratigraphical profile of the site of Gabalovo. It covers the three building horizons from the Middle Bronze Age. The results are presented a quantitative correlation between the studied samples by horizons and by species. The results provide the opportunity to some extent to undertake a palaeo-ecological characteristic. Due to its wide distribution the oak has been a dominant species in the forest vegetation composition which also included elm-tree, maple, birch, hazel tree. By the fences and in the suburbs of the settlement dominant were species as the mountain ash, cornel-tree, cheery, plum and most probably part of those plants were objects of gathering.

Analogical example for dispersed wood is the studied stratigraphical profile of the settlement of Madrets (Chalcolithic) as well as that of Dyadovo (Bronze Age).

In the samples of charred wood taken from the central profile of the tell settlement Madrets the oak is the dominant species. And most often found species are ash-tree, maple and from the fruit bearing ones – plum, cornel-tree, etc.

The analysis of charred wood in the Neolithic site Yabalkovo is also an example of dispersed wood. All samples are collected from floor levels. The data mark wide variety of species.

ANALYSIS OF WOOD FROM CONTEXTUAL POINT OF VIEW

On the basis of conducted studies the following picture appeared: the quantity of fragments established in the different structures has different characteristics. Thus for example those found in hearths and pits provide a much bigger number of wooden fragments than those found in the floor levels of the houses except that the degree of fragmentation is different, which depends on the type of activity, the way of depositing and the degree of carbonization? By this context, concerning the species composition, it is almost always feebly represented, which is connected most probably with the single or two fold activity of the defined structure – i.e. in the most cases we have here homogenous representation of the taxons. The quantity of more charred fragments found in the pits could be explained by their way of usage as the pits were often used as cereal storage places and after such use the pits were burnt and then refilled.

By way of a differenece to than this type of contexts is the heterogeneous representation typified by the study of charred wood mainly by squares, floor levels (Yabalkovo) and stratigraphic columns (Madretz, Galabovo). It provides much more information due to the wider scope of the activity – done first, on a wider space and then, to longer duration of inhabitance.

ANALYSIS OF THE WOOD FROM A TAPHONOMIC POINT OF VIEW

The example given above with Pit No 1 from Adata could be referred to as Class A – where the connection between the context and the plant remains is very close and the in-

formation is reliable.

The samples of dispersed wood could be referred to as Class C – namely that the material appears to be a result of several different activities connected with burning. This material is mixed with archaeological elements – stones, ceramic fragments, daub, etc. The remains here have a feeble connection with the context in which they were found but as a whole they are very important as they provide information about their entire special distribution in the site.

* * *

The list of the wood vegetation is of about 25 different species. On the basis of the adducted data the dominance of oak in most of the studied settlementswas established. Hazel tree, elm-tree and alder were also often found. The studied wooden fragments provide useful information about their use. An example is the wonderfully preserved bases of stakes in the site Orlitza which provide valuable data about the type of wood used in houses construction. A big part of the wood has been from trees of a similar age that were cut at a single time from one and the same place. Most of wood proved to be from oak but from the dwelling floor originated also numerous fragments from ash-tree obtained through flotation. The ash-tree wood is extremely robust and very suitable for carpenter's tools.

A similar example for usage of wood is the wooden objects found in Topolnitsa – nails. One of them was made of oak and the other from Rosacae wood – from plum, pear or similar fruit bearing tree. In the samples from layer No 36 of the same site was found a wooden construction where the wood was oak.

Among the archaeological finds in the site Adata several small silver objects – from sq. B 31/ B 32 were found. In their hollows was preserved charred wood. The analysis of the four studied samples proved the presence of maple.

The charred wood provides useful information for the study of the natural environment of man in different archaeological periods. The use man has made of wood has been a constant process. It is natural to gather wood from the close vicinity and thus we obtain the environmental information. The settlements were located mostly at altitudes of 0 – 500 m, i.e. in the belt of the mixed oak forests by river terraces so it is natural the oak species were dominant. In the mixed forest composition elm-tree, maple, ash-tree and hazel-tree were also included. The pine tree was also found at an altitude 0 – 400 or 500 m but in limestone lands.

As a whole the charred wood data provide evidence for wide use of oak wood in the Neolithic with the presence of alder-tree, elm-tree, ash-tree and fruit-bearing trees by rivers. The data about the pine show its wider distribution in the past.

Chapter 9

Composition of Filling in the Different Structures

— **Pits**
— **Vessels**
— **Pithoi**
— **Houses, floors**

Different ways of storage were used depending on the expected duration of storage and the final goal. The storage may have been silos, pits, granaries, vessels, pithos. Each one of them is oriented towards a certain function. In that connection the characteristics of each one of these negative structures is to be commented below.

PITS

A considerable part of these contexts were pits. The practice of grain storage in dug pits is widely spread in the past. According to ethnographic data the possibility of conservation and the duration of use of a granary do not exceed 20 years. Gransar (1994, pp.401-412) establishes that if the sediments are gravel or sandy they could function from 1 to 5 years while if they were dug in normal soil – they functionred considerably longer.

The ancient sources contain numerous descriptions of such pits – granaries. Plinius for example describes that it is a usual practice in Spain as well as in the other parts of the Mediterranean lands. The storage in underground pits continues to be a common practice beyond this period.

Thus for example Currid and Navon (2008) communicate the availability of pits in Malta, dated from 17th century. Similar structures are found in Sicily from 12th – 15th century. Bresc (1979, pp.113-122) tells of such in France – 19th century. Contemporary examples of such storage are known in many places. Such practice is known in Cyprus and Palestine. The pits are about 2 m deep, usually used for grain or legume storage and they are located in the vicinity of the settlement. According to Currid and Navon (2008, pp. 67-78) physical evidence and historical documents show that in the Iron Age in Palestine in all kinds of soils and loci the pits are commonly spread. The authors

have studied the technique of construction of series of pits from several archaeological sites in the country side Lahav- tell Halif – in Palestine and thus they come to the following conclusions: "...Two standard types of pits are differentiated in the Iron Age in Palestine: cylindrical and bottle-shaped. Their sizes vary from 1 to 2 m width and 2 – 3 m depth. "There exist different concepts about the use of the pits but two stand out as basic. The bottle-shaped pits were used for grain storage, meat and corn whereas the cylindrical ones were for garbage Currid and Navon (2008, pp. 67-78).

According to Cauvin (1994, p.64) the storage pits were spread on a mass scale in the late Neolithic in Greece. The storage is necessary not only for preservation of cereals during the winter but also for a "secondary economy" in regions with seasonal contrasts. Granaries are known in the Near East where the earliest of them is dated in the pre-ceramic Neolithic period A Cauvin (1994, p.64). Most of the traditional vessels typical for the Mediterranean region are finished with clay grout or straw-woven Kanafari-Zahar (1994). The storage places were used not only for conservation of products during the winter but often also the surplus production was preserved for elimination the risk in places where the annual yield could vary significantly.

In that connection I consider that most of those pits found in the pre-historical settlements in the territory of Bulgaria have utilitarian character. With some reservation it could be assumed that the negative structures studied by me could also have been granaries.

TAPHONOMIC ANALYSIS

The results of the taphonomic analysis have shown what could be expected by the study of these structures.

— In the most part in the pits studied we did not established the presence of any considerable quantity of cereals or of other plant products. That is it had either decayed or completely burnt.

— Often we found only single grains and fragments of charred wood.
— The conservation of these seeds and grains was poor. The grains were often deformed and broken.
— The presence of weeds was also poor as they are lighter, smaller and they burn faster.
— Significant fragmentation of wood was noticed.

There may be a number of reasons for these facts but if we suppose that after each use of the pit it has been burnt the aim may have benn to eliminate the activity of different bacteria, insects, and microorganisms.

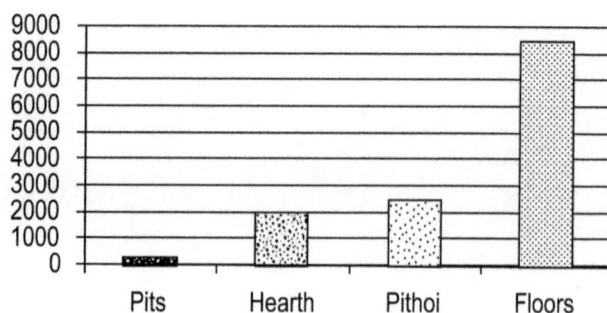

Fig. 40. Quantitative distribution of the plants remains in different negative contexts of the studied settlements.

VESSELS

They also present an example of way of grain storage. In the rare cases when their contents are preserved, they could provide valuable information:

— About the ways of storage making and of cleaning of grain;
— If those are supplies;
— Are they liable to additional processing – often the cereals get storage together with the glumes;
— Are there weeds and of which species;
— Conglutinated grain masses – evidence of presence of water and later a fast charring.

PITHOSES

In some of the settlements I studied pithoses were found. They are often used for keeping water.

HOUSES AND FLOORS

These archaeological contexts are the most reliable as from them could be extracted information about different activities:

— They are results of longer time period.
— The diversity of plant remains is significant though they are not well represented in any great quantity.
— The observations usually show plant remains concentration around ceramic compositions, ovens, hearts and house floors.
— The material in most cases is fragmented as a consequence of its re-depositing due to unintentional activities and it has a heterogeneous distribution.
— The information obtained from these archaeological contexts provides a more comprehensive picture for a longer time period and a defined variation of the different types of wild growing and weed species which have entered into the storage by different farming activities – either in the houses or in the vicinity of the settlement.
— On the other hand the presence of the wood flora is also well documented. Its usage here is connected with different activities – on one hand for household needs and on the other hand for elaboration of different tools, for roof construction, fences and other building elements.

Chapter 10

AN ATTEMPT AT PALAEOECOLOGICAL RECONSTRUCTION

The natural vegetation is a basic factor in the formation of the natural environment. The palaeographic development in the Pliocene and in the Quaternary has influenced the character and the geographical distribution of the vegetation in Bulgaria. Substantial changes in the vegetation took place in the Quaternary when the northern parts of the European continent and the mountains were covered by glaciers. At the end of the Holocene the environmental changes are already connected with the economical activity of man. Mass scale deforestation and grazing started as result of which vast forest spaces were cleared of the trees. Thus the character of the flora was changed by replacement of the natural vegetation with cultivated species.

The information obtained through the pollen analysis is of substantial importance in establishing the anthropogenic processes and to what extent man influenced the surrounding vegetation. The data from the pollen diagrams document the presence of cereals grains as well as such which serve as anthropogenic indicators. They are the ones that explicitly prove the presence of man activity. Examples are the pollen grains of *Cerealia- type, Triticum - type, Hordeum - type, Avena - type* and *Secale*. In addition, there also exist the species called "secondary indicators" such as *Plantago lanceolata, Rumex, Scleranthus, Urtica, Juniperus*, which prove man's presence and more specially the practice of cattle breeding Bojilova, et al., (1990, pp. 48-57).

The beginning of the Holocene after 11 500 cal. years is characterized by a fast increase in the average annual temperatures of 5-6° C. The climatic improvement is expressed with the establishment of a series of termophyles species. The most distinctly expressed climatic oscillations of lowering of the temperature in the Holocene are registered around 9900 - 8200, 5500, 4200, 2500 cal. years and between 1200 and 1650 A.C.

The basic reasons for these temperature variations are the periodic cycles of 1500 (± 500 years), called forth by changes in the currents of the Northern Atlantic Ocean Maslin, et al., (2005).

Such climatic anomalies between 8247-8086 B.P. are discussed in connection with the process of transition from transitional farming and in relation to the geographical dynamics and the trajectories of the populations in South-eastern Europe from Budja (2007). According to his findings anomalies are observed around 8200 B.P. in the palaeo-cilimatic archives in almost global scope. It appears that it should have been colder in a big part of the northern hemisphere during these anomalies which are documented by the shifting of the glaciers and the accelerated atmosphere circulation over the Northern Atlantic and Siberian lands. This climatic event of 8200 cal B.P. was noticed for first time in the registrations of the ice core in Greenland – the results were obtained by the European project for the Greenland ice core – European Greenland Ice Core Project – GRIP.

According to Weninger, et al., (2006) around 8200 cal. year B.P. an abrupt climatic change occurred which caused a drying-up in the western parts of the eastern Mediterranean which in its turn brought the fast spread of the Neolithic culture to South-eastern Europe. In the valley of Struma river a number of Neolithic settlements came into being and they facilitated the successfull development of farming. A substantial change in the composition of the vegetation in South-western Bulgaria is established in the Climatic Optimum of the Holocene when the general tendency towards warming reached its culmination. The Climatic Optimum was preceded by an abrupt freeze around 8200 cal. year B.P. in the northern hemisphere with a duration of 160 years.

In South-western Bulgaria after 8000 cal. year B.P. the spread of coniferous forests with dominant pine-tree and fir-tree which removed the birch communities and partly those of the oak, commenced. The results of series of studies show that by the beginning of the Holocene both the winter and summer anomalies have been positive for South-eastern Europe. About 8000 cal. year B.P. a deep minimum by the summer temperatures – a decrease with 1,5°C was registered, while the winter temperatures remained almost without change. In consequence the climate has changed towards milder winters and cooler summers (Tonkov, et all., 2008).

The developed map for the beginning of the Climatic Optimum of the Holocene (9000-7000 C14 B.P.) in the Mediterranean region including a reconstruction of the vegetation cover reflects wide spread of mixed oak forests by lower altitude in different parts of the Balkan Peninsula (Jalut, et all., 2005). In the region of South-eastern Bulgaria (the Mountains of Rila, Northern Pirin and the Western Rhodopes) in the beginning of the Climatic Optimum by average altitude were developed mixed oak forests with *Tilia, Ulmus, Acer, Fraxinus, Carpinus betulus*, and in the partly deforested lands with *Corylus*. Over them were developed forests composed by birch and groups of *Pinus sylvestris*. To the north of the Balkan mountains vast spaces were occupied by mixed oak forests with participation of birch, lime-tree and hazel while the coniferous – pine, fir tree and juniper had limited distribution (Tonkov, et all., 2008).

The last stage where human presence is clearly manifested in the pollen diagrams is connected with the sudden expansion of the juniper – 2800/2700 cal. year B.P. through the Sub Atlantic. In many places the juniper formed the upper boundary of the forest. The destructive processes started with the cutting down of the coniferous forests and the wide spread of the beech-tree. The expansion of the beech-tree communities was supported by the economy activity of man through both cutting down and burning the coniferous woods for wooden material and for widening the fields for seasonal grazing (Tonkov, et all., 2008).

According to the data of Tonkov obtained from the pollen diagrams from South-western Bulgaria, the following picture emerges:

"...Two periods with presence of anthropogenic indicators get demarcated. The first one is with duration from 10 000/9 300 cal. year B.P. (8000/7300 cal. year B.C.) to around 6500/5600 cal. year B.P. (4500-3600 cal. year B.C.), where these indicators have sporadic participation. In the second period - after 4300/4000 cal. year B.P. (2300 -2000 cal. year B.C) their presence became constant and increased. Between those two periods there is hiatus of 2000-1600 years. In the Iron Age the presence of the anthropogenic indicators had considerably increased. After 2600 cal. year B.P. (600 cal. year B.C.) in Osogovska Mountain the expansion of the beech-tree commenced around 2200 cal. year B.P. (200 cal. year B.C.) and the beginning of the permanent presence of rye is connected with its growing after 1300 AD.

From the results of pollen data intensified anthropogenic presence in the region of Popovi livadi (Sothern Pirin Mountain) in the period around 2530-1300 cal. B.P. or 530-700 A.D. was established, when a significant part of the coniferous and broad leaved forest vegetation was destroyed. Thus in Belassitsa Mountain most substantial changes occurred around 1760 cal. year B.P. - III century AD (Tonkov, et all., 2008).

Our data for charred wood from the studied settlements are a confirmation of the distribution of the oak forests in the lower mountainous belt during the Climatic Optimum of the Holocene. The fragments of *Rosaceae, Corylus, Cornus mas, Carpinus, Acer, Ulmus* often are numerous. The wood of *Pinus* most often belongs to the species *Pinus nigra* and *Pinus syllvestris*. The wood of *Salix, Alnus, Populus* provides evidence that these species have been an integral part of growing plant communities along the riverside that were also used by man. In the determined wood from the archaeological sites Kamenska Chuka and Koprivlen - *Abies, Pinus sylvestris* and *Juniperus* are present, which is confirmed by the palinological data about the spread of coniferous forests of *Abies* and *Pinus* during the Sub-boreal.

With the setting in of the anthropogenic changes the presence of the forest and grass vegetation considerably decreases and the spread of weeds increases.

In the antracologically studied settlements the presence of charred wood is different. It is quite difficult to complete these evaluations – on one hand in connection with the settlements location and their physic-geographical region and, on the other hand for the quantity of the find and studied material.

Thus for example our data for Thrace show domination of oak wood during the studied periods with a considerable increase being observed in the Bronze Age. This is obviously connected to the clearing of the land for grazing and growing cultivated plants.

These data correlate well to the pollen analyses from the area of the tell settlement Ezero and of those from around the Straldja marsh-land. The data of the pollen analysis from the area of the tell settlement Ezero show a relatively low percentage of tree vegetation in the late BronzeAge. This vegetation has been characterized with opened landscape areas and some lands covered by xero-mezophyle oak forests whose percentage participation is about 20-30%. Probable species in such forests were different oak species (*Quercus*), elm-tree (*Ulmus*) and ash-tree (*Fraxinus*). Additionally, the analysis of charred wood from late Bronze Age layers in Ezero additionally shows that those species have been used by the earlier inhabitants. The pollen shows the domination of oak (*Quercus*) with participation of *Ulmus, Fraxinus, Corylus colurna, Carpinus orientalis* and *Abies*, which most probably developed around the Sveti Iliya hills. The moisture-loving vegetation around the lake demonstrates series of changes. For the last 3000 years several different macro-fossil ensembles were established. Thus for the Late Bronze Age is stated characteristic vegetation for shallow-water waters such as *Cyperus fuscus, Zanichellia palustris*, which prove that in the Late Bronze Age the lake has had seasonal fluctuations and even daily cycles. This flora proves the existence of dry periods in summers. The water fluctuation changed during the first half of the Early Iron Age between 1100-900 B.C. Small indications of pollen show that different local springs or streams have been in existence, which maintained more humid climate in the region. Another important moment is the presence of pollen of cereals and the high percentage of *Artemisia, Centaurea, Polygonum aviculare, Rumex acetosella, Compositae*. In conclusion the period Late Bronze Age – Early Iron Age is characterized by distinctive continuation of intensive processing of land. It has been proven is that the grazing areas for animals were more important than the arable land.

According to Chapman, et all., (2009) the vegetation history in the region of Straldja shows that in the period around 4200 cal. B.P. the natural vegetation has been destroyed but small plant communities existed by the Bronze Age settlements. The vegetation was characterized by moisture loving meadow and halophytic grassy communities.

Dominant among the halophytic communities were species of *Chenopodiacae* found mostly in the period between 2830-1330 cal. B.P. when the waters subsided. Partial regeneration of the mixed oak forests was found in the hills around after 1330 cal. B.P. As a result of pollen data the strongest human impact is established in the late Bronze Age and before the Middle Ages.

The archaeobotanical evidences underlines the existence of a well developed economy in this part of the country.

The palaeo vegetation of Straldja correlates with the studies of the peat-bog of Sadovo, located near to the river of Maritza. The age of the studied profile of the bog is similar by radio-carbonic dating to that of Stralja. The pollen diagram of Sadovo shows presence of dispersed communities of *Quercus, Ulmus* and *Tillia*, which have existed since Sub-boreal and have been consecutively destroyed by the local population in the region. Alongside the rivers communities of willow, alder tree and poplar were growing but grassy vegetation was predominant in the region. The palinological and archaeobotanical data prove intensely use of land as arable Tonkov, et all, (2008).

The forests of elm-tree and hornbeam were destroyed as the settlements climbed the mountain slopes. Many ruderal plants appear evidence of intensive cattle breeding in the Bronze Age. By the end of the Bronze Age in the pollen diagrams from the high mountains in Bulgaria a considerable presence of wheat, barley and rye is marked. The expansion of *Artemisia, Plantago lanceolata, Rumex, Cirsium type* is connected with land clearing for grazing territories Bojilova and Popova (1998).

As the economical conditions were extremely favorable it inevitably led to demolition of the coniferous forests and to expansion of beech-tree communities. In the high parts of the mountains in the Bronze Age seasonal grazing was practiced and thus the *Pinus mugo* woods often were burnt out.

Almost identical anthropogenous processes are established for the region of the lake Srebarna and of Varna. The presence of the anthropogenic indicators of the pollen diagrams proves the intensive grazing. The expansion of the hazel (*Corylus avellana*) and the hornbeam (*Carpinus betulus*) could be explained with the mass cutting down of the oak woods by the Chalcolithic population and with the appearance of secondary plants communities Lazarova (1995, pp. 47-67).

The culmination of the Climatic Optimum (from the end of 8000 B.C to the end of 5000 B.C., calibrated C14) is connected with favorable conditions which brought increase in the human population with a result in intensive land processing – evidenced in the settlements of Malak Preslavetz, Durankulak and others, studied achaeobotanically.

The increased presence of cultivated cereals and of anthropogenic indicators such as: *Plantago lanceolata, Plantago media, Plantago major, Polygonum aviculare, Centaurea cyanus* is established in the Bronze Age - uncalibrated 5000-3500 B.P. (3000-1500 B.C.) and in the Early Iron Age which is connected with the Sub-Atlantic period – 3000-2500 B.P. (1200 B.C.) in the pollen diagram from the lake of Srebarna Lazarova (1995).

As result of the reconstruction, though not complete due to the absence of enough information, the following conclusions could be drawn:

1. In all the three studied regions there was ascertained the presence of oak which is a constituent element in the leaf-fall broad-leaved forests. It has been commonly used being easily accessible as it is found in the low lands.

The analysis of the charred wood has shown that the oak has been the dominant tree species.

The hornbeam is also often found as it followed the moderate continental climate.

2. Coniferous: The percentage of wood found from coniferous trees is extremely small and in almost every case the documented species is the Austrian pine tree which was growing in the lower parts in lime lands. Nowadays it has almost disappeared. Charred wood of: *Abies sp.*; *Juniperus sp.*; *Pinus sylvestris; Pinus nigra* is documented in the sites Kamenska Chuka, Koprivlen, Galabovo.

3. The following species were identified and are typical representatives of the vegetation in the river valleys: elm tree (*Ulmus sp.*), alder tree (*Alnus glutinosa/incana*), ash-tree (*Fraxinus sp.*), poplar (*Populus*) and representatives of *Pomoideae* family.

Certain formations of green-fences have existed with the participation of some fruit-bearing trees (plums, cherries, and cornel-tree) as well as ash-trees and elm-trees. The alder tree and the hazel were found around the cleaned spaces and alongside rivers. Most of the tree species are fond of sun shine and humidity. The elm tree has inhabited the lower parts of the woods where there was higher humidity. It requires more moderate-continental climate as well as richer soils. The elm-tree (*Ulmus sp.*) enters into the composition of the broad-leaved mesophile mixed forests and grows alongside rivers.

4. The vegetation in the Valley of Struma River is characterized as Mediterranean type vegetation. The typical species of the region are: *Quecus coccifera, Pistacia sp., Jasminum oficinale, Quercus frainetto, Carpinus betulus, Ulmus campestre.*

The charred wood documented from this region provides evidence of the dominance of oak forests. The species most often found are: *Quercus dalechamii*; *Querus polycarpa.*

The intensive use of the oak forests in the following epochs leads to series of anthropogenic changes connected with the clearing of the forest aiming at increased usage of lands for growing plants; clearing for grazing lands which is connected to the development of cattle breeding.

Chapter 11

CONCLUSIONS

BASIC CONCLUSIONS FOR THE PRE-HISTORIC FARMING

One of the basic questions in studying the pre-historical settlements is: 'what was their environment?' In this respect the settlements archaeobotanically studied by my coleagues and myself provide information about the reconstruction of the palaeoenvironment, the economy and the different way of human subsistence as well as to specify the systems of ancient farming.

The present archaeobotanical study aims at summarizing the basic plants food resources of the ancient inhabitants of the Bulgarian lands during the studied periods applying different types of methods and analysis.

The data from the systematization of the findings from the 36 pre-historic settlements that were studied provide evidence that the basic types of corn in the neolith settlements were hulled wheat – einkorn, emmer, and barley. In most of the studied settlements those wheat are found in big quantities – often as food supplies also in the granaries. Similarly, their presence was found also in many archaeological sites in other areas of the Balkan Peninsula: Franhthi, Ahileon, Sitagri II – Greece van Zeist (1980), Kroll (1983), Obre, Anza – I (Renfrew 1979), Opovo - Yugoslavia Borojevic (1989, 2006), Valamoti (2004).

The cultural plants which were found came from Anatolia and took their place in the local human diet. Obviously the hulled wheat has played a significant role in the sown fields.

Einkorn found in some of the studied pre-historic settlements in Bulgaria is distinguished by coarse grain, different in form from that found in the territory of Asia Minor, which is considered for its basic centre of origin. Some of the grains of naked barley which were found also have difference in morphology. These data provide grounds to assume that the Bulgarian territory appears to be very ancient and, to a certain extent, a self-dependent centre of

development of agriculture. It is possible that the einkorn has been cultivated here independently from Asia Minor as there are traces of wild *Triticum boeoticum* Boiss. in archaeological finds from the Neolithic. It is rather possible that in the past the areal of this wild species, considered as the immediate progenitor of the cultivated einkorn, has covered the territory of Bulgaria. It is also quite possible that this species is known to the south of the Balkans. Anyway, if it has not been directly cultivated here, then at least certain local populations have been established.

A conclusion could be drawn on this basis that the Greek territories and the Balkans have been in close contacts with the Anatolian centre.

The summarized data of Perles (2004) for a number of neolith settlements in Greece also confirmed that the wheat species dominate in the corn ensembles. From the mostly presented einkorn and emmer the dominant one is the emmer.

Our data from the Bulgarian studies of the 36 settlements provide evidence of almost equal presence of these two wheat – the emmer has about 43% of all grain finds in the Neolithic settlements and almost the same percentage keeps also the einkorn – 45%.

The data about *Triticum aestivum* as free-threshing is that it is extremely rare in the Neolithic settlements in Greece Perles (2004). Halstead (2000) also reports that it requires certain soil conditions. Similar are also the results from the studied by us locally found materials – it is found quite rarely in the Neolithic and in Chalcolithic and it starts to be more and more found in the Bronze Age. Quite strange seem the results of Marinova (2005) who reported certain permanent presence of it in more of 10 (?) of the Neolithic settlements in the territory of the country and especially in northern Bulgaria.

Barley has been traced during all pre-historical periods. Initially its hulled form was cultivated. Evidence for it is provided in many of the finds in Neolithic settlements in Greece as well as in the territory of Bulgaria. As a percentage the barley sown fields were probably less or almost the same as those of wheat.

All legumes were present even as early as the Neolithic. In principal , they adept to different ecological conditions.

Thus for example in most of the settlements located alongside Struma River and in the studied settlements of Northern Bulgaria bigger quantities of peas and grass pea are found as they need systematical watering and most probably the conditions there were more favorable for their growing. Whilst in Thrace (in particular in the studied prehistoric settlements from the region of 'Maritsa Iztok') quite often the bitter vetch has been the preferred culture.

All leguminous species found in the territory of Bulgaria have a common distinctive feature – small-sized grains. That is a typical feature for the Mediterranean through which their origin is proven and that their distribution is centralized in this area. The coarse-grained forms appear quite lately – in the Roman epoch.

The ascertainment of chick-peas in the archaeobotanical data proves to be interesting. The chick-peas species is a member of the cereals complex of plants presenting the Near East neolith. Chick-peas is included in the traditional farming in the Mediterranean lands. It has been found in several pre-historic settlements in the territory of Bulgaria.

The finds of chick-peas in countries neighboring to Bulgaria are scanty – in Dimini, 3500 B.C. Kroll (1979), Otzaki – 6000 B.C. – Greece Kroll (1981).

In a big part of the pre-historic settlements the wild plants were a parallel source of subsistence. The wide spread evidence of the species in the vicinity of the settlements provided opportunity for their easy gathering. Recently the finds show 11 species of fruit-bearing trees and 18 species of bushes. Predominant among them are the following species: acorns, hazelnuts, cherries, plums, mountain ash, blackberry, apple, pear, hawthorn, Cornelian cherry, common elder and blue elderberry, blackberry, raspberry.

The data from gathering show the usage of three basic species: Cornelian cherry, grapes, common elder and blue elderberry.

The presence of the Cornelian cherry is followed during the whole pre-historic period and its highest percentage in the settlements is during the neolithic.

Differences concerning the cultivated and the wild growing grapes are observed quite well. In all Neolithic and Chalcolithic settlements the wild grapes have the preva-

lence while in the Bronze Age the characteristic features of cultivation already appear.

The study results provide plenty of information about weeds and wild growing grass vegetation. 92 different species were found, comprising weeds and grass vegetation. These species have wide ecological gamut and they are met in different natural phytocenoses. Some of them have entered the territories of the Balkans and particularly of Bulgaria in the beginning of the Neolithic.

An attempt has been made to classify the sowings based on the character of their accompanying weeds. In that connection the data show that in most of the studied settlements winter weeds were found. Part of those weeds was from the winter crops and thus the existence of winter sowings could be assumed. It is possible also that there existed winter sowings for the crops and spring sowings for vegetables and legumes in the places with water. If an analogy is to be done with some of the studied Greek settlements, then after Perles (2001), Barker and Gamble (1985) the cereals were sown there also in the autumn but spring sowing along the rivers was practiced as well.

On the other hand the found weeds could help in determining definite farming regimes of the past.

The results of my studies significant amounts of information about the weeds and the wild growing grassy plants. Thus 92 different species have been defined, including weeds and grassy plants. The most often found are: *Chenopodium album, Polygonum aviculare, Polygonum lapathifolium, Polygonum persicaria, Rumex acetosa, Rumex acetosella, Bromus secalinus, Agrosthema gitthago, Setaria viridis, Setaria glauca, Vicia tetrasperma, Vicia sativa, Vicia sp., Centaurea sp.* From all of them the following evidenced highest presence: *Chenopodium album, Polygonum aviculare, Rumex acetosa, Rumex acetosella.* The number of the studied weeds in the different settlements, a factor beyond my control. Thus while the establishment of the sowing species in such cases is not always representative, still an attempt to classify the sowings according to the character of the weed has been done. This classification is not always exact as while there are sowings typical for spring or for winter some weed keep an intermediate position. In that connection the data show that in the bigger part of the studied settlement winter/winter-spring weeds were established. Part of the weed species grow in the winter cereal sowings thus supporting the supposition of the existence of winter crop sowings. On the other hand part of weeds are characteristic also of the early spring sowings so that it is difficult to make a really accurate assessment. It is possible that there existed winter sowings for the wheat and spring sowings around the places abundant with waters for the leguminous plants. Still most of the evidence gravitates towards the predominance of winter sowings. Such are the results of the weed flora from the settlements studied in

CONCLUSIONS

the region of 'Maritsa –Iztok' confirmed from the data of Marinova (Marinova, 2002).

From phytho-sociological point of view one part of the weed ensembles is divided into two basic groups: *Secalinetea* and *Chenopodiatea*. The first group characterizes mainly winter (wheat) sowings while the second one characterises – the row crops and leguminous species (Gones, 1992). In the studied settlements representatives of both groups were noticed.

A bigger and more representative sample of the wild growing flora would be needed to provide us with more complete information to establish with greater degree of certainty the regimes of cultivation.

Very often except for food, the wild growing plants were used for different purposes. The data from many settlements provide evidence of intentional gathering of wild species.

Many of the plants were also used as animal fodder. The seeds of some species such as: *Amarantus sp.* - amaranths, *Setaria italica, Setaria glauca, Setaria viridis* – different types of bristle-grass*, Vicia sativa*- common vetch*, Vicia angustifolia* - winter vetch, served as poultry food. Except the previously mentioned species also other ones as: clover, tare, vetch - *Trifolium sp., Vicia sp.* are known as good forage for the animals.

The wild growing plants were often used for different purposes. The data from numerous settlements show elements of intentional gathering. Some authors as Gones (1992), Valamoti (2004) consider some wild growing plants to be typical weed species but they, according to Perles, (2004, pp.25-26) could be also used as food. Such are: *Gallium, Pistacia, Stellaria*, which are eatable. Many of the wild plants need some processing before eating. Such are some species of rye-grass, the bitter vetch, the grass pea which, as also some acorns, are toxic and need soaking in water before being used as food. Many shelled fruits, seeds and fruit stones could be found in the settlement so as many of the plants established in the samples have been used also as animal fodder. The seeds of some species such as for example: *Amarantus sp., Setaria italica, Setaria glauca, Setaria viridis, Vicia sativa, Vicia angustifolia* – different types of rye-grass were used as food for the poultry. Except those species as: clover, rye-grass are good fodders for the animal - *Trifolium sp., Vicia sp.*

Pieces of information about the gathering and the usage of eatable plants and fruits in Greece prove that the list of eatable early Neolithic seeds comprises several species. Seeds of fig-tree are presented in Otzaki, Sesklo, and Argissa Magula. The grapes seeds are sporadically represented and there lack definite indications that the grapes were cultivated – Renfrew (1979). Fruits such as *Prunus,*

Pistacia, Quercus, and Crataegus are found in one of three contexts and in small quantities. The wild fruits are accidentally gathered but Perles (2004) considers that there are no indications for regular consummation with the exception of the fig-tree.

Halstead (1981, p.315) also reckons that the use of wild growing plants by itself is limited mostly to the alluvial basins but it also depends on the distribution, the density of the population, of the yield and of the seasonality. A study of the corn ensembles in Greece definitely underlines a non-regular presentation of wild growing plant resources and even less special technical processes in Greece. The early Neolithic villages in Greece should have eaten much less wild plants that those in Western Europe. All those fruits as acorns, cornel-tree fruit, hazelnut, which are found in high quantities in Europe, here are quire scanty.

In that connection most of the wild growing species in the early Neolithic settlements inhabit opened and dry lands Perles (2004).

Similar is the picture also in the territory of Bulgaria. Heliophilous species as the hazel (*Corylus avellana*) have inhabited the edges of the forests and there they have been noticed and consecutively exploited. Thus the hazel, the cornel-tree and some lianas as the Traveler's Joy (*Clematis vitalba*) the clematis which have elastically wood used for weaving fences and as basis for coating of walls.

The increased use of oak wood during the following epochs led to a series of anthropogenic changes connected with the clearance of forests for expanding the land for cultivation and for pasture grounds, connected with the development of cattle-breeding.

The appearance of such referral plants as *Artemisia, Plantago lanceolata, Rumex, Cirsium type* is connected with clearing lands for grazing grounds.

The floristic content of the finds of plant remains from the different archaeological sites is extremely rich.

The results show wide bio-diversity. Defined are 201 plant species. From them 25 are wild growing trees, 11 species are fruit-bearing and 18 are bushes. The wild growing grassy and weed vegetation number 92 species and the cultivated (cultural) plants are 55.

On the basis of the studied materials some new for the prehistory of Bulgaria plant species were established. These are: *Cicer arietinum and Pinus pinea.*

The genetic resources of the cultural plants appear a valuable fund that, with the development of the contemporary farming more and more irrevocably disappear. In last time increasingly more attention is given to ancient local species and populations.

The tracing of the history and the migration of these ancient cultures as well as the study of the local species obtained through the so-called 'popular selection' could not only provide us with a complete picture of the genetic diversity but also to make it possible to find interesting genotypes in their quality of donors which, by their valuable economic features to preserve the experience of the ancient farmers.

REFERENCES

Alexandrov, St.

1995 The early Bronze age in Western Bulgaria. Periodization and cultural definition. In D.Bailey, I.Panajotov, eds. *Prehistoric Bulgaria*. Monographs in World Archaeology No.22, pp.253-271. Prehistory press, Madison.

2003 Rannata bronzova epoxa v jugo-zapadna Bulgaria - problemi na proučvaneto (Early Bronze Age in South west Bulgaria.The problems of the research). Compendium in memory of P.Gorbanov, *Studia archaeological* suppl.1, pp. 50- 62. Sofia, Universitetsko izdatelstvo "Sv.Kl.Ohridski".

Algaze, G., et al.

1995 Titris Höyük, a small EBA Urban Center. *Anatolica* 21:13-63.

2000 Research at Titris Höyük in Southeastern Turkey: the 1999 season. *Anatolica* 27: 23-106.

Angelova, I.

1982a Razkopki na praistoričeskoto selichte Podgoritza (Excavation of the prehistoric settlement Podgoritza). *Archaeological report from 1981*:11-12.

1982b Razkopki na selichnata mogila Omurtag (Excavation of the tell settlement Omurtag). *Archaeological report from 1981*:16 -17.

1998 Ein Neolitisches Haus aus der localität "Rezervata" bei dem Dorf Drinovo bez. Popovo. In M. Stefanovich et al., eds. *James Harvey Gaul. In Memoriam. In the Steps of James Harvey Gaul*. The James Harvey Gaul Foundation. Sofia. Vol. I, pp. 91-96.

Arnaudov, N.

1937/38 Izledvane na botaničeski materiali ot Sadoveckite razkopki (Analisys of plants materials from the excavations near Sadovo). *Annuaire de L'Université de Sofia, Faculté de Fhisico- chemie*, XXXIV, 3:33-51.

1940/41 Varhu novootkritite rastitelni ostanki ot jujna Bulgaria (About new plant remains from South Bulgaria). *Annuaire de L'Université de Sofia, Faculté de Fhisico- chemie* XXXIV, 3:17-29.

Baczynska, B., and M. Litynska-Zajac

2005 Application of *Lithospermum officinale* in early Bronze Age medicine. *Vegetation history Archaeoboany* 14:77-80.

Bakels, C.

1978 Four Linearbandkeramik Settlements and their Environment: a Palaeoecological Study of Sittard, Stein, Elsloo and Hienheim. *Analecta Praehistorica Leidensia* 11:244-248.

Bar Yosef, O., and A. Belfer-Cohen

2002 Facing environmental crisis, Societal and cultural changes at the transition from the Younger Dryas to the Holocene. In R.T.J.Cappers, S. Bottema eds. *The dawn of farming in the Near East*. Studies in Near Eastern Production, Subsistence and Environment 6. pp. 55-6. Berlin, exOriente.

Barker, M., and G. Gamnble

1985 Beyond domestication: a strategy for investigating the process and consequence of social complexity. In G. Barker, C. Camble eds. *Beyond domestication in Prehistoric Europe*, pp.1-31. Academic Press, London.

Behre, K-E.

1977 Cereals from Neolith settlement Sava near Varna (Bulgaria). *Bulletin de museum* 13 (27): 214 – 215.

Behre K-E., and S. Jacomet

1991 The ecological interpretation of archaeobotanical data. In van Zeist, K. Wasylikova and K-E- Behre, eds. *Progress in Old World. Paleoethnobotany*. pp. 81-108. Rotterdam, A.A. Balkema.

Bittmann, F., and D.Kucan
2004 Erste archäobotanische Untersuchungen in Okoliљte 2002. *Artikel vom 24. Januar 2004* Seite 3 www. jungstein SITE.de

Blumler, M.
2002 Changing paradigms, wild cereal ecology and the origins of agriculture. In R. T. J Cappers, and S. Bottema, eds. *The dawn of farming in the Near East.* Studies in Near Eastern Production, Subsistence and Environment 6, pp. 95-112. Berlin, ex oriente.

Bogard, A.
2004a *Neolithic farming in Central Europe. An archaeobotanical study of crop husbandry practices.* Rutledge, London and New York.

2004b The nature of early farming in central and south - east Europe. *Documenta praehistorica* XXXI: 49-58.

Bojadjiev, J.
2003 Za njakoi problemi na xronologiata i periodizaciata na bronzovata epoxa ot teritiriata na Balgaria (About some problems in chronologie and periodization of Bronze Age in the territory of Bulgaria). Compendium in honor of P. Gorbanov, *Studia archaeological* suppl. 1, pp. 20 – 26. Sofia, Universitetsko izdatelstvo "Sv.Kl.Ohridski".

2004 Kasno xalkolitno selichte Orlitza. (Late Chalkolithic settlement Orlitza). *Archaeological report from 2003:* 56-58.

2006 The role of absolute chronology in clarifying the Neolithization of the eastern half of the Balkan Peninsula. In I.Gatsov I, H. Schwarzberg, eds. *Aegean-Marmara-Black Sea: the present state of research on the early Neolithic.* Schriften des Zentrums für Archäologie und Kulturgeschichte des Schwarzmeerraumes 5, pp.7-17. Radiocarbon. Bronk Ramsey C (1995).

Bojadjiev, J., and K.Bojadjiev
2008 Xalkolitno selichte Varxari, obchtina Momchilgrad (Chalcolithic settlement Varxari, district Momchilgrad). *Archaeological report from 2007:* 51-52.

Bojilova, E.
1986 *Palaeocological conditions and changes of vegetation in Eastern and Southwestern Bulgaria during the last 15000 years.* Doctor of sciences Thesis.

Bojilova, E., and E.Chakalova
1980 Rastitelni materiali ot selichnata mogila pri Djadovo (Plants remains from tell settlement Djadovo). *Expedicio Pontica* 1:155-162.

Bojilova, E., and I. Ivanov
1985 Ekologichni uslovia na Varnenskoto ezero prez neolita i bronzovata epoha spored palonologichni, paleobotanicni i acheolgichni danni (Ecological conditions of Varna lake from Neolithic and Bronze Age after palinological, palaeobotanical and archaeological data). *Izvesia na Narodnija muzei Varna* 21, (36):43 – 48.

Bojilova, E., and Tz. Popova
1989/1990 The role of Balkan Peninsula as a linkage between Asia Minor and Middle Europe in the spreading of Early Agriculture. *Godichnik na Sofiskia Universitet, Biologicheski facultet* 83(2):17-27.

Bojilova, E., S.Tonkov, and D. Pavlova
1990 Pollen and Plant macrofossil analyses of the Lake Sucho Ezero in South Rila Mountain. *Godichnik na Sofiskia Universitet, Biologicheski facultet* 80 (2): 48-57.

Bojkova, A., and P. Delev
2002 *Koprivlen. Rescue Archaeological Investigations along the Gotse Delchev-Drama Road 1998-1999.* Road Executive Agency. Archaeological Institute, Bulgarian Academy of Sciences. Vol.1 Part III: 83-91.

Bökonjy, S., R. Braidword, R. and C. Road
1973 Earliest animal domestication. *Science.* 14 December, vol.162, No 4117:1161-1162.

Bökonjy, S.
1987 *Animal domestication and early animal husbandry in Central, East and South Europe.* BAR, IS .349, pp. 163-169.

Borojević, Ks.
1988 *The relation Among farming Practices, Landownership and Social Stratification in the Balkan Period.* Ph. D. Washington University in St Luis.

2006 *Terra and Silva in the Panonian Plain. Opovo agro-gathering in the Late Neolithic.* BAR, IS. 1563.

Bottema, S.
1974 *Late quaternary vegetation history of Northern Greece.* Thesis, Groningen.

Bresc, H.
1979 Fosses a grains en Sicile (XII –XV siècles) In M. Gast and F. Sigaut eds. *Les techniques de conservation des graines a long Terme I,* pp.113-122. C.N.R.S., Paris.

Budja, M.
2007 The 8200 cal B.P. "climate event" and the process

REFERENCES

of neolithisation in the South Eastern Europe. *Documenta Prahistorica* XXXIV: 191-201.

Cappers, R., R.Neel, & P. Bekker
2009 *Digital atlas of economic plants.* www.plantalas.eu/ea.php-15k

Cauvin, J.
1994 Naissance des divinités, naissance de l'agriculture. In K. Renfrew ed. *La révolution des symboles au néolithique.* C.N.R.S., pp.172-174. Flammarion, Paris.

Chakalova, E., and E. Bojilova
1981 Rastitelni ostanki ot selichnata mogila do gr. Rakitovo (Plants remains of tell settlement Rakitovo). *Intedisziplinarni izledvania* VII-VIII:77-86.

Chakalova, E., and E. Sarbinska
1986 Pflanzenreste aus der neolithischen Siedlung Kremenik bei Slatino. Bez. Kjustendil. *Studia Praehistorica* 8:156–159.

Chakalova, E., and E. Bojilova
2002 Paleoekologicnhi i paleoetnobotanicni materiali ot selichnata mogila do grad Rakitovo. (Palaeoecological and palaeothnobotacial materials of tell settlement Rakitovo). In A. Raduncheva, ed. *Neolithic settlement Rakitovo.* Razkopki i prouchvania XXIX:191-202.

Charles, M., G. Gones, and G.Hodson
1997 FIBS in archaeobotany: Functional Interpretation of weed floras in Relation to husbandry practices. *Journal of Archaeological Science* 24:1151-1161.

Chapman, J. E., E. Magyary, E., and B. Gaydarska
2009 Contrasting subsistence strategies in the early Iron age- new results from the Alfold plain Hungary, and the Thracian plain Bulgaria. *Oxford Journal of archaeology* 28(2): 155-187.

Chohadjiev, S.
2000 Za finala na rannia neolit v basejna na r. Struma (About the end of the early Neolithic in the basin of River Struma). *Starini* 1:61 – 71.

2001 *Vaksevo – praistoricheski selichta* (Vaxcevo – prehistorical settlements).Veliko Tarnovo.

2006 *Slatino– praistoricheski selichta.* (Slatino – pre-historical settlements). Veliko Tarnovo.

Chohadjiev, S., and N. Elenski
2002 Novi prouchvania na selichnata mogila pri s. Hotniza, Veliko- Tarnovsko. (New studies of tell settlement near the village of Hotniza, Veliko Tarnovo

district). Compendium in memory of P. Gorbanov. *Studia archaeological* suppl.1. pp. 24-29. Sofia, Universitetsko izdatelstvo Sv.Kl.Ohridski".

Clark, J. (editor)
1952 *Prehistoric Europe: the economic basis.* London: Rutledge.

Constantini, L.
1981 Semi i carboni del mesolitico e neolitico della Grotta del Uzzo, Trapini. Grotta del'Uzzo. *Quaternaria* 23: 233-247.

Currid, J., and A. Navon
2008 Iron age pits and the Lahav (Tell Halif). Grain storage project. *Bulletin of American schools of Oriental research* 273: 67-78.

Davis, B,. S. Brewer, J. Stevenson, et all.
2003 Data Contributors, the temperature of Europe during the Holocene reconstructed from pollen data. *Quaternary Science Reviews* 22:1701-1716.

Dennell, R.
1974 The purity of Prehistoric Crops. *Proceeding of the Prehistoric Society* 40:132-135.

1978 *Early farming in South Bulgaria from the VI to the III Millennia B.C.* BAR, IS (suplem.) 47. Oxford.

1979 Zemedelski kulturi. (Agriculture). In H.Todorova, ed. *Ezero-early Bronze Age settlement.* pp. 415-425, Bulgarian academy of Science, Sofia.

1983 *European economic Prehistory: a new approach.* London Academic Press.

Docheva, E.
1992 Rastitelni ostanki ot ranno neolitni jilichte v Slatina. (Plant macrorest research of Early Neolithic dwelling in Slatina). *Studia praehistorica* 10: 86-90.

Fagan, B., and C. Decorse (editors)
2003 *In the beginning.* Lavoisier.

Figeral, I.
1990 *Le nord –ouest du Portugal et les modifications de l'écosystème, du bronze final a l'époque romaine, d'après l'anthracoanalyse de sites archéologiques.* Thèse. Université Montpellier II.

1992 Méthodes en anthracologie: étude des sites du Bronze Final et de l'Age du Fer du nord-ouest du Portugal, *Bull. Soc. Fr.,139, Actuel. Bot.* (2, 3, 4): 321- 362.

Franke, G. & K. Hammer (editors).
1977 *Frücte der Erde*. Leipzig, Jena, Berlin Urania – Verlag.

Galabov, J. (editor)
1982 *Geografia na Bulgaria* (Geography of Bulgaria). Sofia.

Ganecovski, G.
2008 Archeologicheski razkopki na rannoneolitno selichte v m. "Valoga" (Dolnite laki) kraj Ohoden, obchina Vraza. (Archaeological excavation from early Neolithic settlement in the countryside "Valoga" (Dolnite laki),Vraza). *Archaeological report from 2007*:30-35.

Gebel, H.G.
2004 There was no centre: The polycentric evolution of the Near Eastern Neolithic. *Neo-lithics* 1/04:28–32.

Georgiev, G.
1974 Stratigrafia i periodizacia na neolita i xalkolita v dnechnite Balgarski zemi. (Stratigrafie and periodization of Neolithic and Chalcolithic in the nowadays Bulgarian lands). *Archeologia* 4:1-18.

Georgieva, D.
1984 Geologia i proekt za exploatacionno prouchvane na jujnata chast na uchastak "Centralen rudnik - Trajanovo-I". Mariza-Iztok. (*Geology and project for exploitation in the south part of "Centralen rudnik – Trajanovo I, Mariza Iztok"*). These bacalaur: Minen – Geological institute. Sofia.

Georgieva, P.
1999 Early Eneolithic Pottery from a Burnt Dwelling – Kozarewa Mogila, a Tell near Kableshkovo. In M. Stefanovich et al., eds. *James Harvey Gaul. In Memoriam. In the Steps of James Harvey Gaul*. The James Harvey Gaul Foundation. Vol. I, pp. 415-425. Sofia.

Gones, G.
1992 Weed phytosociology and crop Husbandry: identifying a contrast between ancient and modern practice. *Review of Palaeobotany and Palynology* 73:133-143.

Görsdorf J, and J.Bojadziev
1996 Zur absoluten Chronologie der bulgarischen Urgeschichte. Berliner C14 Datierungen von bulgarischen archäologischen Fundplätzen. *Eurasia Antiqua* 2:105–173.

Gransar, F.
1994 Le stokage alimentaire sur les etablissments ruraux de l''age de fer en France septentrionale: complémentarité des structures et tendances évolutives. *Actes de colloque du P.C.R., Les établissements agricole de l'Age du Fer dans le nord de la France. E.N.S. 29-30 novembre*, pp. 401- 412. Paris.

Gregus, P. (editor)
1955 *Identification of living Gymnosperms on the basis of xylotomy*. Akademiai Kiado. Budapest.

1959 *Holzanatomie der Europaischen laubholzer ind Straucher*. Akademiai Kiado. Budapest.

Gross, E., S.Jacomet, and J.Shibler
1990 Stand und Ziele der wirtschfsarchäologishen Forschung an neolithischen ufer-und inselsiedlungen im unteren Zürichseeraum (Kt. Zurich,Szhweitz) In J. Shibler, ed. *Festschrift Hans R. Stampfli Shibler*. Verlag Helbing and Lichtenhahn. Basel, 77-100 *Documenta praehistorica* XXXI: 49-58.

Gurova, M.
1999 Trasologicheski analis na ranno-neolitni kremachni kolekcii ot Trakia, Jujna Bulgaria. (Us – wear analysis of early Neolithic flints assemblages in Thrace, South Bulgaria). *Starini* 2: 59-77.

Hajnalova, E.
1975 Rastitelni nahodki ot selichnata mogila pri Goljamo Delchevo. (Plants remains of tell settlement Goljamo Delchevo). In H.Todorova, ed. *Excavation and research*. Sofia.Vol.5: 303-314.

1980 Paleoetnobotanicheskie nahodki iz mnogoslojnovo Novozagorskovo poselenia. *Studia praehistorica* 4: 91-98.

Halstead, P.
1981 Counting cheep in Neolithic and Bronze age. In I. Hodder, G. Isaac, eds. *Pattern in the Past* . Hammond Studies of honor of David Klarke, pp. 307-339.

1995 Plough and Power: The economic and social significance of cultivation with the Ox-drawn Ard in the Mediterranean. *Bulletin of Sumerian agriculture* 8:11-22.

2000 Land use in postglacial Greece: cultural causes and environmental effects. In P. Halted and C. Frederick, eds. *Landscape and Land Use in postglacial Greece*, pp.110 – 128. Sheffield, Academic Press Edmonds.

REFERENCES

Hansen, J. M.
1978 The earliest seed remains from Greece: Paleolithic through Neolithic at Franchthi Cave. *Ber.Dtch. Bot. Ges.* 91:39-46.

Harlan, J. (editor)
1987 *Les plantes cultivée de l'homme.* Press. Univ. France.

Heintz, K.
1990 Dynamiques des végétations holocène en Mediterranée Nord occidentale d'après l'anthracoanalyse des sites préhistoriques: *Methodologie & Paléoécologie, Paléoécologie continentale* Vol. XVI, N. 2, Montpellier.

Helbaek, H.
1959 Domestication of food plants in the Old World. *Science* 130:365-372.

1960 Ancient crops in the Shahrzoor valley in Iraqi Kurdistan. *Summer* 16:79-81.

1964 First impressions of the Çatal Hüyük plant husbandry. *Anatolian studies* 14:121-12.

1969 Plant collecting, dry-farming and irrigation agriculture in prehistoric Deh- Luran. In F.Hole, K.V. Flannery, and J.Aneely, eds. *Prehistory and human ecology of the Deh- Luran Plain*, Memoirs Mus. Anthrop. No. 1, pp. 383-426. University of Michiga, Ann Arbor.

1970 Plant husbandry of Hacilar. In J.Mellaart, ed. *Excavation at Hacilar*. Vol.1 pp. 189-244. Edinburgh University press.

Herre, and M. Röhrs
1977 Zoological consideration on the Origins of farming and domestication. In C.A. Reed, ed. *Origin of agriculture*. pp. 245-279. Mouton, The Hague.

Hilmann, G.
1975 The plants remains from tell Abu Hureyra: A preliminary report. *Proc. Prehist. Soc.* 41: 70-73.

1981 Reconstructing crop husbandry practices from charred remains of crops. In R. Mercer, ed. *Farming practices in British prehistory*, pp.123-162. Edinburgh University press, Edinburgh.

1984 Interpretation of archaeological plant remains: the application of ethnographic models from Turkey. In van Zeist and W. A. Casparie, eds. *Plants and Ancient man*, pp.1-42. A. A. Balkema, Rotterdam.

2000 Plant food economy of Abu Hureyra. In A.Moore, G.

Hillman and T. Legge, eds. *Village on the Euphrate. From foraging to farming at Abu Hureyra*, pp. 372–392. Oxford, University Press, New York.

Hillman, G.C. and M. S. Davies.
1992 Domestication rates in wild wheat and barley under primitive cultivation. In P.C. Anderson, ed. *Préhistoire de l'agriculture: nouvelles approches expérimentales et ethnographiques*. Monographie du CRA No. 6, pp.111-160. Paris, France.

Hopf, M.
1961 *Pflanzenfunde aus Lerna/Argolis.* Zuchrer 31:239-247.

1962 Bericht über die Untersuchungen von Samen und Holzkohleresten von der Argissa – Magula aus den präkeramischen bis mittelbronze- zeitlichen Schichten. In V.Milojčić, J. Boessneck and M. Hopf, eds. *Die deutschen Ausgrabungen auf der Argissa – Magula in Thessalien*. Vol.1.pp.101-110. Bon.

1973 Frühe Kulturoflanzen aus Bulgarien. *Jahrb. Röm. German. Zentralmus*. Mainz.20:1-47.

1974 Pflanzenreste aus Siedlungen der Vinča- Kultur in Jugoslavien. *Jahrb. Röm. German. Zentralmus.* Mainz. 21: pp.1-11.

1983 Jericho plants remains. In K. M. Kenyon and T.A. Holland, eds. *Excavations at Jericho*.Vol. 5. pp. 576-621. British school of Archaeology in Jerusalem, London.

1988 Früneolithische Kulturpfhlanzen aus Poljanica Plateau bei Targovishte (Bulgarien). *Studia praehistorica*. 9:34-36.

Hubbard, R.N.
1980 Development of Agriculture in Europe and the Near east: Evidence from Quantitative Studies. *Economic Botany* 34:51-67.

Ivanov, I.
1983 Spasitelni razkopki na rannoeneoltno selichte pri Suvorovo, Varnenski okrag. (Rescue excavation from early chalkolithic settlement near Suvorovo, district Varna). *Archaeological report from 1982*:21-22.

Jacomet, S., C.Brombacher, and M.Dick
1989 Archäobotanik am Zürichsee. Ackerban Sammelwirtschaft und Umwelt von neolithischen und bronzezeitlichen Seufersiedlungen im Raum Zürich, Zürich, pp.128-144.Orell Füssli Verlag.

Jalut, G. J. Carrion, F. David, S.Tonkov et all.

2005 The vegetation around the Mediterranean basin during the last glacial maximum and the Holocene Climatic Optimum. *The Mediterranean basin: The last two climatic extremes. Explanatory notes of the maps scale 1/5000000).*

Januchevicth, Z.

1991 Nahodki kulturnix rastenii iz pozdnoneoliticheskix sloev Ovcharovo. In H.Todorova, ed. *Ovharovo.* IX, pp. 106-125. BAN, Sofia.

Jakar, J.

1991 *Prehistoric Anatolia. The Neolithic transformation and Early Chalcolithic Period.* Tel Aviv. Monograph series, number 10.

1985 *The Later Prehistory of Anatolia. The Late Chalcolithic and Early Bronze Age.* BAR 268,

Jones, G.

1998 Distinguishing food from fodder in the archaeobotanical record. *Environmental archaeology* : 95-98.

Jones, G., and P. Halstead

1995 Maslins, mixtures and monocrops: on the interpretation of archaeobotanical crop samples of heterogeneous composition. *Journal of archaeological Science.* 22:103-114.

Kanafani-Zahar, A. (editor)

2004 *Liban. Le vivre ensemble* (Hsoun, 1994-2000). Paris, Geuthner.

Ken – ichi – Tanno, G., and G. Willcox

2006 The origins of cultivation of *Cicer arietinum* L. and *Vicia faba* L. early finds from Tell el Kerkh, northwest Syria, late 10 millennium B.P. *Vegetation history and archaeobotany* vol. 15, N 3:197-204.

Knorzer, K-H.

1973 Der bandkeramishe Siedlugsplatz Langweiler 2: Pflanzliche Grossreste. Rhein. *Ausgarb.* 13:139-152.

Koukouli-Chryssanthaki Ch., and H. Todorova, et all.

2007 Promachon-Topolnitsa. A Greek-Bulgarian archaeological project. In H. Todorova, M. Stefanovich, et all., eds. *The Struma/Strymon River Valley in Prehistory. In the Steps of James Harvey Gaul*, Gerda Henkel Stiftung. Sofia,Vol. 2.

Kroll, H.

1979 Kulturpflanzen aus Dimini. *Archaeo-Physika* 8:173-189.

1981 Thessalishe Kulturpflanzen. *Zetrscr. Archäeol.* 15: 67-103.

1983 Kastanas. Ausggabungen in einem Siedlungshugel der Bronze – und Eisenzeit Makedoniens 1975-1979. Die Pflanzenfunde. *Prahistorishe Archaologie Sudosteuropa.* vol. 2. pp.1-176. Verlag Volkes Spiess, Berlin.

1991 Südosteuropa. In W.van Zeist, K.Wasylikova et all., eds. *Progress in Old World.* Palaeoethnobotany. pp.161-177. A. A.Balkema, Rotterdam, Brootfield.

Kruk, J.

1980 *The Neolithic settlements of south Poland.* BAR, IS, 93.

Küster, H.

1991 Phytosociology and archaeobotany. In D.R. Harris and K.D. Thomas, eds. *Modeling ecological change,* pp.1-26. Institute of archaeology, University College. London.

Lazarova, M.

1995 Human impact on the natural vegetation in the region of lake Srebarna and mire Garvan (northeast Bulgaria)- palynological and palaeothnobotanical evidence. In E.Bojilova and S. Tonkov, eds. *Advance in Holocene palaeocology in Bulgaria.* pp. 47-67. Pensoft Publ, Sofia – Moscow.

Leshtakov, K.

1992 Izledvania varhu bronzovata epoha v Trakia. Sravnitelna stratigrafia na selichnite mogili prez Rannata Bronzova epoha v Jgo-iztochna Bulgaria. (Study of Bronze Age in Thrace. Comparative stratigraphy of Early Bronze Age tells settlements in South – East Bulgaria). *Annuaire de L'Université de Sofia, Faculte d'Istoire* 84/85:68 -73.

2002 Some Suggestions Regarding the Formation of the "Thracian Religion". *RPRP* 5:19-51.

Leshtakov, K., B.Borisov, and Tz. Popova

2005 Nadgrobna mogila II (Goljama Detelina) do s. Goljama Detelina, obchina Radnevo. (Grave tumulus II. Goljama Detelina, near Goljama Detelina municipality of Radnevo). *"Mariza-Iztok"- Archeologicheski prouchvania* III:65-87.

Leshtakov K., T. Kancheva-Russeva and S. Stoyanov

2001 Prehistoric Sites. Settlement Studies. *Maritsa-Iztok. Archeologicheski prouchvania* vol. 5:32-48.

Leshtakov, K., et all.

2007 Preliminary Report on the Salvage Archaeological Excavations at the Early Neolithic Site Jabalkovo in the Maritza Valley-2000-2005 Field Seasons. *Anatolica* vol. 33:185-234.

REFERENCES

Lichardus, M.
2000 Mission archéologique de la vallee du Strimon. Fouilles néolithiques Franco-bulgares de Kovacevo. *Rapp. 14.* Campagne de 1999.

Lisicina, L., and L.Filipovich
1980 Paleoetnobotanicheskie nahodki na Balkanskom poluostrove. *Studia praehistorica,* 4: 5 – 90.

Marinova, E.,
1999 Analiz na botanicheski macroostanki ot ranno i kasno neoliten material. (Plants remains analysis of early and late Neolithic material). In. V.Nikolov, ed. *Tel Kapitan Dimitrievo. Razkopki:1989-1999.* pp.123-130. Sofia –Pechtera.

2001 *Vergleichende palaeoethnobotanische Untersuchung zur Vegetationsgeschichte und zur Entwicklung der prahistorischen Landnutzung im Bulgarien.* Dissertation Bonn University.

2002a Archeobotanicno izledvane na neolitnoto zemedelie v dnechna Bulgaria. (Archaeobotanical study of Neolithic agriculture in Bulgaria). *Archeologia* 2:13-24.

2002b Mittel – und spatneolithische botanische Funde. - In S.Hiller & V.Nikolov, et all., eds. *Karanovo. Die Ausgrabungen in O 19.* Salzburg – Sofia, Bd. II:171-179.

Marinova, E., et all.
2002c Ergebuise archäobotanischer Unterzuchungen aus dem Neolithikum und Chalkolithicum in Südwesbulgariun. *Archaologica bulgarica* 3:1-11.

Marinova, E.
2003 *Paleoethobotanical study of Early Bronze Age II in the Upper Stryama Valley (Dubene – Sarovka IIB).* BAR, IS 1139, vol. 2: 499-504.

Marinova, E., and Tz. Popova
2008 *Cicer arietinum* (chick pea) in the Neolithic and Chalcolithic of Bulgaria: implications for cultural contacts with the neighboring regions. *Vegetation history and Archaebotany* vol.17, suplem.1:73-801.

Marinval, Ph.
2001 L'archeobotanique. *Supplément du #382 d'Archéologie* p. 108.

Maslin, G., et all.
2005 *Stable isotopes in paleoclimatology.*

Nesbitt, M. and D. Samuel
1996 From staple crop to extinction. The archaeology and history of hulled wheats. Taxonomy, evolution, distribution and origin. In S. Padulosi, K.Hammer and J. Heller, eds. *Hulled wheats. Promoting to conservation and use of underutilized and neglected crops,* proceeding of the First International Workshops on Hulled Whreats, 21-22 July, 1995. Castelvechio Pascoli. Tuscany. Italy. International Plant Genetic Resources Institute, Rome. Italy. 4, pp.41-99.

Nikolov, B.
1992 *Periodizacia na neolotnite kulturi v Severna Bulgaria – ot Jantra do Timok.* (Periodization of Neolithic culture in North-west Bulgaria from Jantra to Timok). *Izvestia na muzeite v Severo-Zapadna Bulgaria* 18:1-11.

Nikolov, V.
1995 Dve jilichta i keramichnite im kompleksi ot plasta Karanovo III selichnata mogila Karanovo. (Two dwellings and pottery complex from layer Karanovo III in the tell settlement Karanovo). *Archeologia* 4:19-26.

1998 *Prouchvania varhu neolitnata keramika v Trakia. Keramichite kompleksi Karanovo II-III, III i III-IV v konteksta na Severo-Zapadna Anatolia i Jgo- iztochna Evropa.* (Study of neolithic ceramic in Thrace. Ceramic complexes Karanovo II-III, III and III-IV in the context of North-West Anatolia and South – East Europe). pp. 48-49; 16-17. Agato, Sofia.

1999a Varianti na prexod ot rannia kam kasnia neolit v Trakia dolinata na r. Sruma. (Variations of transition from early to late neolith in Thrace in Struma river valley). *Starini* 1:5-11.

1999b Kulturno-xronologicheski problemi na rannia neolit v dnechna zapadna Bulgaria. (Cultural –chronological problems of early Neolithic in nowadays West Bulgarian lands). *Starini* 2:59-66.

Ovharov, N., and D.Kodjamanova, et all.
2008 Archeologichesko prouchvane na skalno svetilichte pri Tatul, Momchilgradsko. (Archaeological study of a rock sanctuary by the village of Tatul, Momchilgrad district). *Archaeological report from 2007:* 542-549.

Özdugan, M.
2001 In Defining the Neolithic of Central Anatolia. *The Neolithic of Central Anatolia – CANeW Round table.* Istanbul, 23-24 November, pp. 223-261. Istanbul.

Palmer, C.
1988 The role of fodder in the farming system. *Environmental archaeology.* vol.1: 1-10.

Panajotov, I.
1991 Ranna i sredna bronzova epoha v Gorno Trakijskata

nizina. Novi problemi. (Early and Middle Bronze Age in the Upper Thrace. New problems) *"Mariza-Iztok"- Archaeological studies* 1:33-41.

Panajotov, I., I. Gazov, and Tz. Popova
1985 Malak Preslavetz – "Pompena stanzia" – rannoneo-liticheskoe poselenie s intamuralnimi pogrebeniami. *Studia praehistorica* 11-12:51 – 61.

Panajotov, I. and D.Valcheva
1989 Acheologicheskite kulturi ot kasnata bronzova epo-ha v Balgarskite zemi. (Archaeological cultures of the Late Bronze Age in Bulgarian lands). *Vekove* 1: 5-16.

Panajotov, I. et al.
1991 Selichnata mogila Galabovo-kasen halkolit i sredna bronzova epoha. (Tell settlement Galabovo – late Chalcolithic and Middle Bronze Age. *"Mariza-Iztok"- Archelogicheski izledvania* I: 141-153;154-164

Panajotov,I., and S. Alexandrov
1995 Mogilen nekropol ot rannata bronzova epoha v zem-lichata na selata Mednikarovo I Iskriza (Tumulus necropolis from early Bronze age in the neiborhood of Mednikarovo and Iskritza). *"Mariza-Iztok"- Archeologicheski izledvania* III: 87-113.

Panajotov, I.,M. Xrisov, and R.Mikov
2005 Spasitelni archeologicheski prouchvania do s. Jaz-dach, obchina Chirpan. (Rescue archaeological excavations in the Jazdach, Chirpan municipality). *Archaeological report from 2004*:78-79.

Perles, K. (editor)
2004 *The early Neolithic Greece. The first farming com-munities in Europe.* Cambridge University press. Edinburgh Building Cambridge, CB2, 2 RU, UK .

Pernicheva, L.
1973 Prehistoric cultures in the Middle Struma valley: Neolithic and Eneolithic In D. Bailey, I. Panajotov, eds. *Prehistory of Bulgaria.* Monographs in World Archaeology 22, I, pp. 99-128. Madison/Wiscon-sin.

Perkins, D.
1973 The beginnings of animal domestication in the Near East. *American Journal of Archaeology* 77: 279-282

Popova, Tz.
1985 Paleoetnobotanicheskie ostanki bliz poselenie Podgoriza i Omurtag na teritorii Bolgarii. *Izvestia Akademii nauk Moldavskoi SSR. Ser. Bilogicheskix i himicheskix nauk*: 69-70.

1990 Izledvane na ovagleni rastitelni ostanki ot srednove-kovna sgrada Silistra. (Charred plant remains of Middle Age building of Silistra). *Interdisciplinarni izledvania* XVII: 63 – 65.

1991a Paleoetnobotanicheski izledvania v Severo-iztochna Bulgaria. (Palaethnobotanical study of North–east Bulgaria). *Archeologia* 2:49-54.

1991b Palaeoethobotanical investigation in South Bulga-ria. Vila Nova de Famalicao. *Palaeoecologia et Ar-ceologia* II:187-189.

1991c Palaeoethnobotanical study of the Yunatsite - Bron-ze Age Settlement - Pazardzik area, South Bulgaria. *Palaevegetation Development in Europe. Proceed-ing of the Pan-European Palaeobotanical Confer-ence,* Vienna -19 -23 September, pp. 69-72.

1992a Analyse preliminaire de restes carpologiques de Kovačevo. *Mission archeologique de la vallee du Strimon. Fouilles neolithiques Franco-Bulgares de Kovačevo.* Universite de Paris I. Centre de resherches Protohistorique, I, vol. 8 : 26-30.

1992b L'analyse des restes végétaux carbonises du tell Dijadovo, Symposia Thracologica 9. *Bibliotheca Thracologica* 11, pp.238-241. Bucuresti.

1994 Archeobotanichen analiz na selichnata mogila Ma-dretz "Gudgova mogila" i Iskritza. Predvaritelen analiz. (Archaeobotanical analysis of tell settlement Madretz "Gudgova mogila" and Iskritza. Prelimi-nary repport)*"Mariza- Iztok" - Archeologicheski prouchvania.* II:119-121.

1995a Arheobotanichni materiali ot kasnoneolitnoto ji-lichte v selichnata mogila Karanovo. (Archaeobo-tanical remains from late Neolithic house in the tell settlement Karanovo). *Archeologia* 4:27-28.

1995b Palaeothnobotanical remains from the early Bronze Age settlement of Galabovo (South Bulgaria).In H. Kroll, R. Pasternak, eds. *Res archaeobotanicae – Proceeding of 9th International Workgroup for paleoethnobotany Symposium*, pp.261-266. Kiel.

1995c Plants remains from Bulgarian Prehistory (7000-2000 B.C.). - In D. Baily, I. Panajotov, eds. *Prehistory of Bulgaria.* Monographs in World Archaeology, 22, I. pp. 193-207. Madison, Wiskonsin.

1999 Etude Carphologique et anthracologique de tell Kajmenska cuka (Blagoebgrad) – Bronze final.In *"Thracian World at the crossroads of civilization". Procc. of the VII Intern. Congress of Thracology.* Constanza – Mangalia - Tulca, II: 477-481.

REFERENCES

2001b Archaeobotanical studies. - In: *Maritsa-Iztok. Archeologicheski izledvania* vol. 5: 211-219.

2001a Analiz na ovagleni rastitelni ostanki. (Charred plants remains). In S.Chochadjiev, ed. *Vaksevo. Praistoricheski selicha. S prinosi ot V.Genadieva, M. Gurova, Tz.Popova, L. Ninov.* pp.31-32. Veliko Tarnovo.

2003 Paleoetnobotanicheski izledvania v rajona na Koprivlen (Palaethnobotanical studies from the region the Koprivlen). In P. Delev, A. Bojkova, eds. *Koprivlen.* I, *Resue Archaeological Investigations along the Gotse Delchev-Drama Road 1998-1999.* Vol.1 p. 279-289. Road Executive Agency. Archaeological Institut, Bulgarian Academy of Sciences.

2006 Archeobotanicheski izledvania na teritoriata na Severo-iztochna Bulgaria. (Archaeobotanical studies in the territory of North-east Bulgaria). *Helis* V:509-518.

2008 Archeobotanicni materiali ot selichnata mogila Hotniza. (Archaeobotanical materials from the tell settlement Hotnitza). In M.Gurova, ed. *Praistoricheski prouchvania v Bulgaria. Novite predizvikatelstva.* (Pre-historical studies in Bulgaria. The new challenges). Repport from National praehistorical conference: 26-29.04.2006. pp.189-194. BAN, NAIM, Pechtera.

Popova, Tz., and E. Bojilova
1995 Palaeocological and palaeobotanical Data from the Bronze Age in Bulgaria. In M. Stefanovith, H.Todorova, H.Hauptmann, eds. *In the Steps of James Harvey Gaul. In memoriam.* Vol.1, pp. 391-399. Sofia.

Popova, Tz., and E. Marinova
2008 Palaeoetnobotanical data in South-West region of Bulgaria. In H. Todorova, M. Stefanovich, G. Ivanov, eds. *The Struma/Strymon River Valley in Praehistory. In the Steps of James Harvey Gaul.* Vol. 2, pp. 517-526, Gerda Henkel Stiftung. Sofia

Powel, J.
1999 Center for Anatolian Ethnography and Textile Studies in Istanbul. http://www/creativespirits.de/

Renfrew, J.
1966 A report on recent finds of carbonized cereal grains and seeds from prehistoric Thessaly. *Thesalika* 5: 21-36.

1968 A note on the Neolithic grain from Can Hasan. *Anatolian Studies* 18:55-56.

1973 *Palaeoethobotany. The prehistoric food plants of the Near East and Europe.* Methuen. London.

1976 Carbonized seed from Anza. In M. Gimbutas *Neolithic Macedonia as reflected in the excavation of Anza, south-east Yugoslavia.* Monimenta Archaeologica 1, pp. 303-312. University of California, Los Angeles.

1979 The first farmers in South East Europe. *Archaeo-Physika* 8: 243-265.

Renfrew, K. & P.Bah (editors)
1993 *Archaeology. Theories, Methods and Practice.* Thames and Hudson Ltd. London.

Schoch, H., and S.Pawlik et all.(editors)
1988 *Botanical macro-remains.* Bern, Stuttgart Haupt.

Sherrat, A.
1980 Water, soil and seasonality in early cereal cultivation. *World Archaeology* 2: 313-330.

Sherratt, A.
1981 Plough and Pastoralism: Aspects of the secondary products revolution. In I. Hodder, G. Isaac and N. Hammond, eds. *Patterns of the Past: Studies in honor of David Clark.* pp. 261-307. Cambridge University press, Cambridge.

Sigaut, F.
1988 A method for identifying grain storage techniques and its application for European agricultural history. *Tools and tillage* 6:3-32.

Sinskaja, E. (editor)
1969 *Istoricheskaj geograpfia kulturnoj flori.* Kolos. Leningrad.

Stefanova M.
2000 Control Excavations at Mihalich in 1998 - 1999. *Reports on Prehistoric Research Projects (RPRP)* 4:21-31.

Stefanova, I., and B. Ammann
2003 Late glacial and Holocene vegetation belts in the Pirin Mountains (South west Bulgaria). *The Holocene* 13 1: 97-107.

Stefanovich, M., and A. Bankoff
1995 Kamenska Čuka 1993-1995 – Late Bronze Age site in Southwest Bulgaria. Preliminary findings. In M.Stefanovith, H.Todorova, H.Hauptmann, eds. *In the Steps of James Harvey Gaul. In memoriam.* vol. I, pp. 255-339. Sofia.

Sweingruber, F. (editor)
1988 *Microscopic wood anatomy.* Swiss federal institute

of forestry research, CH-8903 Birmensdorf, Zurcher AG, CH-6301 Zug.

Stanev,P. (editor)
2008 *Orlovetz. Neolithen complex*(Orloovetz. Neolithic complex), pp.24-27.Veliko Tarnovo, Faber.

Stefanova, I. & L. Filipovich
1977 Polenov analiz na torficheto Bogdan – 6 v Sredna gora. (Pollen analysis of turf-bog Bogdan - 6 in Sredna Gora). *VII National conference, Biological facultet, University of Sofia "Sv.Kl.Ochridski"*, maj, Sofia, pp.103-116.

Thanheiser, U.
1977 Botanische Funde. - In S.Hiller and V. Nikolov, eds. *Österreichisch-Bulgarische ausgrabungen und forschungen in Karanovo. Archaölogisches Institut der Universität Salzburg, Archaölogisches Institut mit Museum der Bulgarischen Akademie der Wissenchaften*, Bd. I, Kapitel, 22: pp.429-477.

Thery, J., J. Grill, and J. L.Vernet, et all.
1995 First use of charcoal. *Nature*, vol. 373, 6514: 480-481.

Thissen, L.
2007 C CANeW 14C databases and 14C charts southwest and northwest Anatolia, central Anatolia and Cilicia. Accessed on 17.02.2008. http://www.canew.org/download.html

Todorova, H. (editor)
1989 *Durankulak I.* BAN, Sofia.

Todorova H. (editor)
2002 *Die prahistorischen Graberfelder – Durankulak*. II. Sofia. Deutsches Archaologisches Institut.

Toncheva,G.
1973 Praistorichesko nakolno selichte Strachimirovo II (Prehistorical lacustrine settlement Strachimirovo). *Izvesia na narodnia muzej Varna* 9 (XXIV): 285 – 289.

Tonkov, S., E. Bojilova, E. Marinova and H. Junger
2008 History of vegetation and landscape during the last 4000 years in the area of Straldza mire. *Phytologica Balcanica* 14(2):185-191.

van Zeist, W.
1970 The oriental Institute excavations at Mureybut, Syria. Preliminary report on the 1965 campaign. Part III. *Palaeobotany. Journal of Near East Studies.* 29:167-176.

1972 Palaeobotanical results of the 1970 season at Cayönü, Turkey. *Helinium* 7: 3-19.

1974 Palaeobotanical studies of settlement sites in the coastal area of the Netherlands. *Palaeohistoria* 16: 223-271.

1975 Preliminary report on the botany of Gomolava. *Journal of Archaeological Science* 2: 315-325.

1980 Aperçu sur la diffusion des végétaux Cultivée dans le région Mediterranean. Natur. *Monspelensia*, pp.124-145.

1988 Some aspects of early Neolithic plant husbandry in the Near East. *Anatolica* XV: 49-69.

van Zeist, W and Bakker – Heeres
1979 Some economic and ecological aspects of the plant husbandry of Tel Aswad. *Paleoorient*: 161-169.

van Zeist, W and S. Bottemma
1971 Plant husbandry in Neolithic Nea Nikomedia, Greece. *Acta bot. neerl.* 20: 524-538.

Waines, H.
2007 Domestication and crop physiology roots of green wheat. *Annals of botany* 100 (5): 991-998.

Vakarelski, Ch. (editor)
1974 *Ethografia na Bulgaria*. (Etnography of Bulgaria). Sofia, Nauka i izkustvo.

Valamoti, S.
2004 *Plants and People in the late Neolithic and Early Bronze age North Greece. An Archaeobotanical Investigation.* BAR, IS. 1258, Oxford.

2007 Agriculture and use of space at Promachon Topolnitza:Preliminary observation on the archaeobotanical material. In H. Todorova, M. Stefanovich, G. Ivanov, eds. *The Struma/Strymon River Valley in Praehistory. In the Steps of James Harvey Gaul.* Vol. 2, pp. 545-551 Gerda Henkel Stiftung. Sofia

Weninger, L., E.Alram-Stern,. E.Bauer, L.Clare et all.
2006 Climate forcing due to the 8200 cal yr B.P. event observed at Early Neolithic sites in the east Mediterranean. *Quaternary research* 66(3): 401-420.

Vilatias, J.
1992 *Atlas de malas hierbas*. Ediciones Mundi Prens. 2 edision

Willcox, G.
1995 Wild and domestic cereal exploitation: new evidence from early Neolithic sites in the north Levant and south-east Anatolia, *Arh*. Vol.1, N 1: 9-1.

2005 The distribution, natural habitats and availability of

REFERENCES

wild cereals in relation to their domestication in the near East: multiple events, multiple centers. *Veg. Hist. Archaeobot.* 14:534-541.

2006 How fast wild wheat domesticated. *Science* vol. 311:1886.

Willerding, U.
1983 Pälao-Ethobotanik un Ökologie". *Verhandlugen der Gesellshaft für Ökologie (Festhrift Ellenberg).* 11:489-502.

Wilkinson, G., and C. Stevans (editors)
2003 *Environmental Archaeology: Approaches, Techniques and Applications.* Tempus, Stroud.

1972 The wild progenitor and the place of cultivated lentil (*Lens culinaris* Medik.). *Econ. Bot.* 26:326-332.

Zohary, D., M. Hopf (editors)
1988 *Domestication of Plants in the Old World. The origin and spread of cultivated plants in West Asia, Europe and Nile Valley* . Oxford Science Publication, Clarendon Press, Oxford.

2000 *Domestication of Plants in the Old World. The origin and spread of cultivated plants in West Asia, Europe and Nile Valley.* 3rd revised ed. Oxford Science Piblication, Clarendon Press, Oxford.

www.ingramcontent.com/pod-product-compliance
Lightning Source LLC
Chambersburg PA
CBHW061006030426
42334CB00033B/3382